# The
# Social Sex

# The
# Social Sex

——— A HISTORY ———
OF FEMALE FRIENDSHIP

## MARILYN YALOM

*with* THERESA DONOVAN BROWN

HARPER ● PERENNIAL

NEW YORK ● LONDON ● TORONTO ● SYDNEY ● NEW DELHI ● AUCKLAND

HARPER ● PERENNIAL

HarperCollins books may be purchased for educational, business, or sales promotional use. For information please e-mail the Special Markets Department at SPsales@harpercollins.com.

FIRST EDITION

*Designed by Michael Correy*

Library of Congress Cataloging-in-Publication Data

Yalom, Marilyn.
The social sex : a history of female friendship / Marilyn Yalom with Theresa Donovan Brown. — First edition.
pages cm
Includes bibliographical references.
ISBN 978-0-06-226550-0
1. Female friendship. I. Brown, Theresa Donovan. II. Title.
BF575.F66Y35 2015
302.34082—dc23
2015008011

OV/RRD   10 9 8 7 6 5 4 3 2 1

For Irv and Paul

# CONTENTS

# PREFACE

THIS BOOK OWES ITS INCEPTION to the loss of a beloved woman, Diane Middlebrook. When Diane died in December 2007, I lost a friend who had been my colleague, confidante, and sister author for more than three decades. In the months following her death, I painfully realized just how precious and irreplaceable her friendship had been. Others who knew Diane shared this bittersweet perception. Having watched Diane's determined fight against the tumor that eventually devoured her, having seen her find sustenance in writing until the very end, I could think of no better way to honor her memory than to craft a book about female friendship.

We Americans take for granted the ability of women to choose their friends and spend time with them as they wish.

We blithely assume that this is the "natural" state of affairs. Yet in many parts of the world today, girls and women must ask for permission from their fathers, mothers, husbands, brothers, or older sisters to meet with their friends—that is, if they are allowed to have them at all. Even in America, friendships are often subject to parental or spousal control. Parents certainly try to direct their children toward other children whom they consider "suitable," and marriage often means that the new wife has less time for her single friends. Even in America, a woman's freedom to choose and spend time with her friends is circumscribed by familial, economic, and cultural considerations.

The subject of friendship is less glamorous than the subject of love, which still commands center stage in life and literature, not to mention the publishing world. I, too, have been guilty of contributing to the plethora of books focusing on love (e.g., *How the French Invented Love*) without considering its near relative. During the many months of our collaboration, Theresa Donovan Brown and I have explored the overlap between love and friendship and have often found that it is difficult for us to make clear-cut distinctions. What is friendship? Is it so different from love?

Theresa and I, friends as well as collaborators, are keenly aware that we are writing from a specific place and from a specific moment in time. What we have to say about women as friends is undoubtedly colored by our own situations, yet we hope to cast our net beyond our narrow enclave in Northern California and encompass the experiences of many other women elsewhere. Friendship is the birthright of all American women today, and, for many of them, the crown jewel of personal life. Looking back through history, it is remarkable and edifying to see how friendship evolved as a real option for women. Precious and important as these bonds are for anyone who cares about or relies upon women, a history of female friendships cautions us against taking them for granted.

Marilyn Yalom

# The
# Social Sex

D O MEN OR WOMEN HAVE more friends? The answer today in the United States is most likely to be women. Popular wisdom tells us that women are basically more sociable, more open, more empathetic, more nurturing, more collaborative, more "friendly" than men. The media reinforce this stereotype with films, TV shows, and "chick lit" novels that depict the close bonds many girls and women enjoy throughout their lives. Moreover, some scholarly studies have contended that women develop deeper, more intimate friendships than men, and that women's friendships are vital both to their mental health and, in evolutionary terms, to the survival of their offspring.[1] In the case of married couples, if the wife dies first, the husband often becomes isolated, depressed, or physically ill, whereas if the reverse occurs—if he dies first—she is often sustained by her friends.[2] Good friends—whether women or men—are now considered indispensable to the well-being of American women, whatever their age.

The difference between male and female friendship patterns has been a pop-culture and academic hot topic for at least twenty-five years.[3] Most academic studies conclude that man-to-man friendships differ from woman-to-woman friendships. One social scientist put it this way: "Guys get together and have shoulder-to-shoulder relationships—we do things together—as compared with women, who are more apt to have face-to-face relationships."[4] Many women confide in their friends, while many men simply enjoy hanging out together. All too often, competitiveness colors male relationships and prevents men from disclosing their frailties and pain to their friends. Hence men's intimate discussions are often reserved for their girlfriends, spouses, or platonic women friends. This allows many men to project a public persona that is independent and self-sufficient—"masculine" qualities par excellence.

On the other hand, a woman without close friends, however successful she may be, is often judged to be lacking the emotional capital long associated with the female sex. Adolescent girls and college-age women are often thought to rely heavily on their friends for empathy and feedback. While young women may pull back from time-consuming friendships in the initial stages of a marriage, they seem to find friends again as the need arises: they seek out female colleagues and mentors as they negotiate the demands of the workplace, they communicate with other mothers as they raise their children, they exchange worrisome secrets as they undergo menopause or divorce, and they lean on one another as they suffer through cancer, other illnesses, or the death of a spouse. How many times have we heard a woman say: "I would never have made it without my friends."

This prominence of women as friends would have surprised people living in the distant past. Almost all the documents on friendship during the first two thousand years of Western history—from 600 BCE to 1600 CE—pertain to men. Of course, almost all these documents were written by men, for other men. But the focus on men's friendships is more than a question of gendered authorship and readership. Male authors extolled friendship as a male enterprise, necessary not only for personal happiness but also for civic and military solidarity. When an ancient Greek philosopher invoked friendship as the noblest form of human attachment, he did not consider women worthy of attention since they were noncitizens, nonsoldiers, and nonparticipants in the public realm. Sequestered in the domestic sphere of Greek homes, women might have had friends among themselves, but what did that contribute to the greater good?

Moreover, women were notoriously "weaker" than men and were considered constitutionally unsuited for friendship at the highest level, according to a negative view of female friendship that persisted long after the Greeks and Romans. Their rivalries, jealousies, and lack of steadfast loyalty would be thrown in women's faces for centuries to come. Indeed, as late as the mid–nineteenth century, the British *Saturday Review* posed the question of whether women were even capable of friendships within their own sex.[5] The prolific California author Gertrude Atherton would opine in 1902: "The perfect friendship of two men is the deepest and highest sentiment of which the finite mind is capable; women miss the best in life."[6] And C. S. Lewis, author of the *Chronicles of Narnia*, wrote in 1960 that the presence of women in male circles contributes to "the modern disparagement of friendship"; such women should be left to their "endless prattling" and not be allowed to contaminate the superior intercourse of male minds.[7] Even today, films and TV shows emphasize the mean-spirited cliques of catty teenagers and the sexual rivalries of young women, following a tradition of stereotypes that has long undermined the value of women as friends.

The evidence for friendship among Greek and Roman men of the citizen class, among medieval clerics and crusaders, and among Renaissance humanists is substantial. Though separated by space, time, language, and culture, they wrote extensively about the virtues of male friendship in a variety of genres, including letters, treatises, memoirs, and stories. For example, the rousing French epic *The Song of Roland* (circa 1100) features the heroic friendship of Roland and Olivier on the battlefield, following a literary tradition that goes back two millennia to the characters of Achilles and Patroclus in Homer's *Iliad*, and even earlier, to the Babylonian story of Gilgamesh and Enkidu. In contrast, female friendship was not a theme in classical or medieval literature except in rare cases that usually revolved around a heterosexual love affair, with one woman playing the role of confidante to another.

Elsewhere during the Middle Ages, in the quiet confines of Christian monasteries, intimate friendships sprang up among the monks who lived, worked, and prayed together. Venerable leaders and future saints such as Anselm of Canterbury (1033-1109) and Bernard of Clairvaux (1090-1153) wrote numerous letters that expressed deep affection for other men of the cloth, regardless of whether they were superiors in rank, like abbots, priors, or bishops, or simply fellow monks. But by the time of Anselm's death in 1109, comparable letters were being written by women living in nunneries. Those of Hildegard of Bingen (1098–1179), no less crafted in Latin than those of Saint Anselm, testify to the close friendships women experienced in convents. Hildegard's forceful personality shines through the many epistles she sent to several women she knew and loved. Today her correspondence, like that of Saint Anselm, comprises three volumes. But despite the numerous women friends who received and answered her letters, the public face of friendship remained resolutely male.

In a classic example of men's assuming that friendship was exclusively a male province, the renowned Italian humanist Leon Battista Alberti (1404–72) wrote a treatise, "On the Family," in which he imagined the following scene, recounted by a rich Florentine merchant soon after his marriage: "Then she [his wife] and I knelt down and prayed . . . that He might grant us the grace to live together in peace and harmony . . . and that He might grant to me riches, friendship, and honor, and to her, integrity, purity, and the character of a perfect mistress of the household." Whether Alberti intended the merchant's words to be taken seriously or not, he reflected the aspirations of Italian husbands, for whom friendship with other men figured prominently in their daily lives—as opposed to women, who were advised to limit their activities to their families and households.[8]

The sixteenth-century French writer Michel de Montaigne (1533–92) provided the quintessential example of friendship between men. His relatively short but passionate association with Étienne de La Boétie, which was immortalized in one of Montaigne's best-known essays, "Of Friendship," was modeled on the figures of Greek and Latin literature whom the two men had studied together. Attempting to live up to classical ideals, they aspired to a union that would be no less than Aristotle's idea of "one soul in two bodies." After his untimely death, La Boétie found his way into Montaigne's writing as a permanent presence and enduring inspiration. When Montaigne asked himself why he loved La Boétie, his answer evoked the mystery of mutual attraction: "Because it was he, because it was me."

Typical of his and earlier eras, Montaigne's public pronouncements on women as friends were entirely negative. He wrote that "the ordinary capacity of women is inadequate for that communion . . . nor does their soul seem firm enough to endure the strain of so tight and durable a knot."[9] Ironically, at the end of his life, when Montaigne engaged in his most serious friendship since the death of La Boétie, it was with a woman—the young Marie de Gournay, who helped edit the final edition of his *Essays* and devoted herself exclusively to his literary and personal needs.

Given this long history of celebrating men as friends, how

is it that women were able to co-opt the public face of friendship? Surely women in the past always had some sort of female
bonding, even if it went undocumented. When did friendship
among women become visible to the world at large and celebrated as a prized staple of their lives? With the exception of
medieval nuns, European women did not begin to leave written records of their views on friendship until the fifteenth century. Once vernacular languages displaced Latin in writing,
women took up the quill with greater ease and wrote to their
friends with increasing frequency. Some also penned essays
and fiction, so that from the time of Christine de Pizan's *Book
of the City of Ladies*, written in French around 1405, we have
evidence of women's friendships from their own points of view.
In Italy, Moderata Fonte (1555–92) wrote a mini-dialogue on
friendship based on the argument that "women make friends
with other women more easily than is the case with men" and
that such friendships are more lasting.[10]

By the time of Fonte's death in 1592, a new era of friendship
had opened up for women, not only on the continent in France
and Italy but also across the channel in England. There, many
women of the upper and middle classes were acquiring new
freedoms, among them that of bonding publicly with other
women. Shakespeare's plays reflected the newfound alliances
women formed, particularly to protect one another against
misguided men (e.g., Beatrice and Hero in *Much Ado About
Nothing* and Portia and Nerissa in *The Merchant of Venice*).

Subsequently, the salons created by seventeenth-century
French *précieuses* and eighteenth-century English Blue Stock-

ings allowed women citizenship within the most socially elevated friendship circles of their time. Whether they were
single-sex or coed, these circles encouraged members to seek
out potential friends with whom they could then meet in the
privacy of their own bedrooms and parlors. By the end of the
eighteenth century, women's friendships with other women
had become a respected, time-consuming part of their lives—
second only to the care they devoted to their families. It is
true that this model applied mainly to women with means:
peasants were lucky if they found sufficient time to care for
their families, animals, and crops; and working-class women
could not practice the niceties that friendship required among
the better heeled. Working women had to enjoy friendship on
the run, often when necessity called one friend to assist another in childbirth, illness, or death. But for more privileged
women, entertaining female friends was a sign of their social
position, to be savored in private and even trumpeted about
among their peers.

Such a scenario was not limited to Europe. By the time of
the American Revolution, the rituals of friendship had taken
hold throughout the thirteen colonies. Friends in the same
locale visited one another on a regular basis; if their homes
were far apart, letters bridged the distances between them.
The remarkable correspondence between Abigail Adams and
Mercy Otis Warren gives us a detailed picture of the friendship between two exemplary American women.[11] Each identified herself as the wife of a public servant and as the mother
of several children, with numerous domestic responsibilities;

yet each set aside substantial time in order to maintain their friendship, primarily through letters, since they lived at a considerable distance from each other—Abigail in Braintree, Massachusetts, and Mercy in Plymouth.

If 1600 marked the beginning of a grudging social recognition of women's claim to friendship in Europe, 1800 was the turning point that changed the public face of friendship in both Europe and the United States. Increasingly, friendship came to be understood as a feminine, not just a masculine, endeavor. Indeed, it can be argued that the entire concept of friendship, especially in the Anglo-American world, became feminized. Girls and women began to write letters to one another in a language of love not so different from the language of heterosexual longing. Words like *dearest, darling, precious, heart, love,* and *devotion* flowed easily from the pens of Victorian girls and women as they passionately corresponded with one another. The formation of numerous clubs based on religious, ethnic, political, and cultural interests allowed middle-class, as well as upper-class, women an opportunity to meet in social groups that spawned countless pairs of friends. The establishment of girls' schools, seminaries, and colleges (e.g., Mount Holyoke, Vassar, and Wellesley in the northeast United States; Randolph-Macon, Mary Baldwin, and Agnes Scott in the South; and Mills on the West Coast) became breeding grounds for lifelong friendships.

By the nineteenth and twentieth centuries, the idea that friendship was exclusively or even primarily a male affair had been largely reversed. Women were thought to be more caring, more tender, and more loving than men, and thus more suited for friendship. Friendship itself became identified with the feminine characteristic of emotional intimacy and was no longer synonymous with heroic or civic comradeship—though men periodically made attempts to reassert the hegemony of earlier forms of male bonding. Female friendships, once denigrated by men and often experienced by women themselves as little more than the by-products of family relations, became highly valued in their own right. For the past 150 years, the stock in women's friendship has been on the rise.

A search on the Google Ngram site, which tracks the frequency of appearance of particular words and phrases in 5.2 million digitized books covering a time frame from 1500 to 2008, shows a huge upward spike for the phrase "women friendships" in the second half of the nineteenth century, after a flat line for three and a half centuries.[12]

Our book about friendship, covering a period of more than two millennia, does not pretend to be all-encompassing. Yet we approach the topic in the hopes of presenting a familiar subject in a radically new way. We shall look at the evolution of women as friends within certain time frames and within specific cultures, as we believe that friendships cannot be understood without attention to the settings in which they take place. Medieval German nuns, sixteenth-century village "gossips" in England, seventeenth-century aristocrats in France, colonial women in early America, working girls during the Industrial Revolution, pioneer women on the American western frontier, twentieth-century feminists, and twenty-first-century wage earners—each of these groups is buttressed by the social structures that surround them.

By looking at friendship historically, we may be able to grasp why these crucial human bonds—women's friendships—were once marginalized and why they have now, finally, come into ascendency. Why do we care? Because the past is prologue. Because we dwellers on this crowded, conflicted planet must ply every relationship tool available to us. The female friendship model has always been around, but unfortunately the former stewards of our common history tended to ignore it. No more. The power, and often the wisdom, of what women seek and find in friendship could lead future generations into lives of dignity, hope, and peaceful coexistence.

# PART ONE

# When the Public Face of Friendship Was Male

# ONE

# LOOKING FOR FRIENDSHIP
# IN THE BIBLE

*"Thine own friend, and thy father's friend, forsake not."*
—PROV. 27:10 (KING JAMES BIBLE)

*"Greater love hath no man than this, that a man lay down his life for his friends."*
—JOHN 15:13 (KING JAMES BIBLE)

MOST FRIENDSHIP STORIES IN THE Hebrew Bible and the New Testament are male-centered, but often we can sense a feminine presence lurking within the narrative. At a time when men had a monopoly on writing and showed little interest in what women did among themselves, it is surprising that we have any records at all of women as friends. While the headliner male friendship stories are more prominent and more familiar to us, a few portraits of female bonds supplement our understanding of friendship among Biblical peoples.

## The Book of Job

In the Book of Job, the central character is a blessed man who has lived righteously; yet God, goaded by Satan, decides to strip Job of all his possessions and slaughter his children. When this isn't enough, He afflicts Job with boils from head to foot. Why God has done this has perplexed readers for almost three millennia. When Job sits down to lament his fate, three male friends come to comfort him and share his sorrow: "They sat down with him upon the ground, seven days and seven nights."[13] Nobody talks for a whole week. Finally, Job breaks the silence with a famous (and understandable) self-pitying rant, cursing the day he was born. Then a cycle of dramatic interaction begins in which each friend argues with Job and tries to make him admit to having sinned and accept God's chastisement without question. But an anguished Job continues to maintain his innocence, and in so doing he raises eternal questions about the nature of goodness, evil, and divine justice.

Though his friends think they have done their best to sympathize with him, Job calls them "miserable comforters" who do not truly understand his situation: "I also could speak as ye do; if your soul were in my soul's stead." Here is the crux of friendship as experienced not only by Job but by friends in all centuries: Can we truly put ourselves in another's shoes? Can we truly share another person's "soul"? How should we behave when our friend is anguished, depressed, suicidal? Job says he would not criticize his friend, he would not "heap up words against you, and shake mine head at you. But I would strengthen you with my mouth, and the moving of my lips should assuage your grief" (Job 16:2–5). When you are in distress, what you need is a loving presence—someone to hold your hand and empathize with your sorrow, not offer criticism or blame.

In the end, after the appearance of God Himself out of the whirlwind, Job confesses his lack of understanding and acknowledges the sovereignty of God's judgment. His friends have played a significant role in Job's psychological trajectory, if only as a sounding board for his protestations. They are present at Job's final reversal of fortune, when God restores him to his previous state of happiness. We can assume that as good friends, they share in his joy.

### David and Jonathan

Like the story of Job, the story of David dates from the time of the patriarchs, roughly 1000 BCE. As described in the books 1 Samuel and 2 Samuel, the lofty bromance between David and Jonathan offers a paradigm of pure love in which "the soul of Jonathan was knit with the soul of David" (1 Sam. 18:1). When Jonathan's father, Saul, commands his servants to kill David, Jonathan speaks up on David's behalf and protects him from death. Saul wishes to destroy David because he, rather than his own son, is destined to become the king of Israel. But Jonathan cares only about the survival of his friend and tells David: "Go in peace, for as much as we have sworn both of us in the name of the Lord, saying, The Lord be between me and thee, and between my seed and thy seed for ever" (1 Sam. 20:42). Thus Jonathan and David establish an alliance of brotherly love and loyalty that will extend unto their offspring.

### Ruth and Naomi

Women as platonic soul mates, comparable to Jonathan and David, are not visible in the Hebrew Bible. The closest we come to such affection between women is the story of Ruth and her mother-in-law, Naomi. When Ruth is widowed, instead of returning to her own Moabite clan, she chooses to follow Naomi, reciting these now-famous words: "Whither thou goest, I will go; and where thou lodgest, I will lodge; thy people shall be my people, and thy God my God" (Ruth 1:16). Though these women are depicted as capable of faithful attachment to each other, their relationship comes about because they have been connected by a man, in this case Naomi's son, who was Ruth's husband. But whatever the motives that inspire her choice, Ruth's attachment to Naomi comes across as genuine. Here is one woman casting her lot with another woman, through bonds of friendship that resemble those of marriage. It is not surprising that "Whither thou goest . . ." is now recited as part of the vows that some couples—opposite-sex as well as same-sex—pledge to each other in their weddings.

### Plural Wives

Elsewhere in the Hebrew Bible, the husband is often a source of friction rather than bonding between women. Remember the story of Sarah, Abraham's wife, and her maidservant,

Hagar (Gen. 16). Because Sarah is barren, she asks Abraham to give her a child through Hagar's body—long before the technological advances of the twentieth century, the Hebrews had their own form of surrogate mothering. Hagar gives birth to Ishmael, but a jealous Sarah then sends her away. Later, with God's intervention, Sarah does indeed conceive a son, Isaac, but the rivalry between the mothers does not end. It is passed on to the sons, Isaac and Ishmael, who each found separate nations. Isaac continues Abraham's Hebrew lineage, whereas Ishmael is considered the legendary father of the Arab nations.

The story of Sarah and Hagar fits into the stereotypical depiction of women as rivals for male attention, which persists to the present day. This same rivalry is depicted between Rachel and Leah, two sisters given as wives to Jacob (a grandson of Sarah and Abraham). Jacob labors for seven years to marry Rachel, but he is tricked into marrying her elder sister, Leah, instead. Then he is obliged to work seven more years for the woman he had wanted in the first place. After fourteen years of hard work he finds himself with two wives, Leah and Rachel, who are envious of each other for different reasons. Leah envies Rachel because Jacob loves Rachel more, and Rachel envies Leah because Leah has produced children, while Rachel has remained barren. At this point, a contest ensues between the two women over childbearing. Rachel produces a son through her handmaiden and, later, one on her own. Leah produces more sons and, when she is no longer fertile, enlists her handmaiden to add to the number of

Jacob's children. All in all he becomes the father of a minitribe consisting of numerous sons and one daughter, Dinah. Instead of portraying women as sisters and friends, aiding each other in childbirth and nursing, as they must have done, the Biblical author chose instead to depict them as jealous rivals for their husband's affection and contestants in childbearing. Where is the historical truth in all this?

The Hebrew Bible presents childbirth stories in the context of a tribal or national continuum. So important was the idea of perpetuating the Hebrew people that no fault was found if the husband of a childless woman fathered offspring via the wife's handmaidens, with the consent, or even urging, of the wife. But when we come to the New Testament, individual actions and interactions begin to take precedence over tribal considerations. Thus we see more relationships between non-family members and members of different tribes in the New Testament than we do in the Hebrew Bible. The Gospel depictions of Jesus and the Apostles give us the archetype of today's bromances.

### Jesus and His Disciples

Several of the apostles have highly distinctive personalities—Peter, the stalwart leader, dense and cowardly in some moments, brilliant and rock solid in others; Thomas, the stubborn pragmatist; Matthew, the calculating one; John, the most beloved by Jesus (at least according to John himself); Judas, the needy one. Many of these characters have backstories that add heft to their individuality, apart from tribal or familial connections. Most of them are framed as ordinary guys—fishermen, a tax collector, maybe a drifting son of a well-to-do father. The friendships we witness throughout the gospels are, with two exceptions discussed below, man-to-man bonds. The male writers of the New Testament had no qualms about giving short shrift to women in their narratives. Even the Virgin Mary, in a notable scene, is ignored by her then-famous son, who emphasized the primacy of one's chosen associates over the value of family:

> While he yet talked to the people, behold, his mother and his brethren stood without, desiring to speak with him.
>
> Then one said unto him, Behold, thy mother and thy brethren stand without, desiring to speak with thee.
>
> But he answered . . . and he stretched forth his hand toward his disciples, and said, Behold my mother and my brethren!
>
> For whosoever shall do the will of my Father which is in heaven, the same is my brother, and sister, and mother.
>
> *(Matt. 12:46–50)*

Despite the significant differences between the New Testament and the Hebrew Bible, relatively little changes between the two books with regard to women. We do hear mention of the women who supported Jesus's ministry, and, after his death, those who continued as followers within the fledgling church. These women provided funds, food and shelter, and meeting places, but few warranted the barest character development or description of their friendships.

## Elisabeth and Mary

There is, however, one New Testament story that truly centers on female friendship. This is the lovely story of the Visitation, when the young, newly pregnant Mary goes to visit her kin, Elisabeth, who is finally with child after a long history of being barren. One cannot deny women center stage when it comes to pregnancy.

In the brief, elliptical Biblical passages that mention the meeting between the mothers-to-be of John the Baptist and Jesus, the scripture writer sticks to his script and reinforces the virgin-mother-of-God conceit. Elisabeth's advanced-age pregnancy is so outré that the annunciating archangel, Gabriel, tells Mary about it as proof that God can do anything and thus can get Mary pregnant while she remains a virgin.

Chances are good that Mary's visit to Elisabeth was welcomed by Mary's family, who found themselves with the problem of a betrothed virgin turning up pregnant. Joseph of Nazareth would have been firmly within his rights to reject Mary and publicly shame her family. But Joseph, whatever his reasons, opted in favor of acceptance. As soon as Mary conceived, she went "with haste" (Luke 1:39) to the hill country of Judah, where Elisabeth and her husband, Zacharias, lived.

When Mary arrived and "Elisabeth heard the salutation of Mary, the babe leaped in her womb" (Luke 1:41). This action has been interpreted as the joy John the Baptist felt in utero when confronted with the divine fetus of Jesus. Elisabeth and Mary say nothing about how it felt to be two women sharing the discomforts and fears of pregnancy, one already six months along, the other possibly experiencing the morning sickness so common in the early stages. Luke tells us that Mary remained with Elisabeth for three months, which would coincide with Elisabeth's giving birth.

The Elisabeth/Mary story endures. It has reinforced, for hundreds of years, one of Christianity's most basic tenets: the virgin birth of God-made-man. Luke keeps the focus in the Visitation episode on the male fetuses, yet the fundamental power of the meeting between the two pregnant women vibrates just below the surface. The strength of the story lies in a motif to which everybody responds: the bond between two women who love each other as friends. And, since time immemorial, pregnancy and birthing are likely to top the list of occasions when a woman requires, and is permitted, the friendship of another woman.

The actual details of this important story are minimal. Both women clearly needed help. We do not know how old Elisabeth was—at what age a woman was considered hopelessly barren in Biblical times—but we do know that pregnancy can be complicated and difficult for anybody, and especially for older mothers. On the other hand, social custom of the time would indicate that Mary, a virgin, was very young indeed—twelve to fifteen years old. Given her mission, we can easily speculate that this girl was terrified. We can imagine how much comfort the two women provided for each other, with Mary offering the physical strength and cheering aura of youth and Elisabeth responding with the practical and emotional experience of a mature woman.

### Mary Magdalene

There is another standout Mary in the New Testament, Mary Magdalene, and we know even less about her than we do about Jesus's mother. The few sentences referring to "the Magdalene" in the four Gospels only hint at her place in Jesus's ministry. She and two other named women (Joanna and Susanna) mentioned among his followers "ministered unto him of their substance"—that is, provided him with material support (Luke 8:3). At one point, when Mary is afflicted with a mysterious illness, Jesus is able to drive evil spirits from her body. Alongside Jesus's mother and his mother's sister, Mary of Cleophas, Mary Magdalene witnesses the crucifixion (John 19:25), and, on the third day after the crucifixion, she discovers the empty sepulchre where Jesus's body had been placed. Indeed, she is the only person to be listed in all four Gospels as the first to discover Jesus's missing body and to believe that he had been resurrected (Matt. 28:1–10, Mark 16:1–11, Luke 24:1–11, John 20:11–18). Both Mark and John name Mary Magdalene as the one who announced the news of Jesus's resurrection to his disbelieving apostles.

Because of Mary Magdalene's participation in key moments of the crucifixion and resurrection, some scholars and creative writers have speculated that she was more than a bit player in Jesus's life—not only his avid student but possibly also his lover or wife. They have found support in the noncanonical Gospel of Mary Magdalene, extant in Coptic translations from the Greek, which states specifically that Jesus loved Mary more than any other woman.[14] In this vein, Mary

Magdalene has inspired the work of New Testament scholars, most notably Elaine Pagels,[15] as well as novelist Dan Brown in *The Da Vinci Code* and composer Mark Adamo in his beautiful opera *The Gospel of Mary Magdalene*. A case for the Magdalene as Jesus's wife can be made when one asks why Mary, Mary's sister, and Mary Magdalene are the only women who attend Jesus at the cross, and why Mary and Mary Magdalene go together to claim his body at the tomb. It would have been appropriate for Jesus's mother and his wife to fulfill these functions together. Just as we have speculated above on the friendship between Elisabeth and Mary during their pregnancies, it would be equally fitting to imagine the friendship of the two Marys—one Jesus's mother, the other his disciple and possibly his wife. It does not take a great stretch of the imagination to picture them mourning together and trying to comfort each other at the crucifixion.

Similarly, if we cast back in the story to the time of Jesus's ministry, we can imagine the friendships between Mary Magdalene, Joanna the wife of Chuza, and Susanna, all three of whom are identified as having provided Jesus with sustenance. What did they say to one another in their rudimentary kitchens as they prepared the bread and wine that would later become ritualized as the Eucharist? How did they manage to travel together as a trio of female companions, along with the other disciples following Jesus? It was no small matter that Jesus accepted women among his followers. Surely the women struck up friendships with one another, and possibly with the men as well. We can only bemoan the sketchiness of their presence in the Gospels as compared to the gravitas accorded those twelve male disciples, who gave the world the first model of Christian brotherhood.

Because the Bible is the foundation text for Christians and Jews, its passages on friendship still have resonance for millions of people worldwide. Both straight and gay men can look to David and Jonathan as icons of male bonding, spelled out in poetic language that expresses their loving loyalty to one another in the face of life-threatening warfare and murderous intent. Women have no comparable biblical icons. Female friends may be inspired by Ruth and Naomi, but it is essentially a family story. And though women sense a kind of "sisterhood" in the story of Mary and Elisabeth, it focuses primarily on the meeting of the two fetuses. Would that the Bible had told us more about Mary Magdalene, for she comes across as the most meaningful figure for contemporary women. A woman who counted among Jesus's disciples and who was recorded as witness to the crucifixion and resurrection—that woman's friendships beg our imagination. It is not surprising that writers and composers have recently taken up her story and tried to imagine the relationships that would have been possible in early Christian times for a woman such as the Magdalene.

# TWO

# PHILOSOPHERS AND CLERICS

*"Friendship is a single soul dwelling in two bodies."*
*"Perfect friendship is the friendship of men who are good, and alike in virtue."*

—ARISTOTLE, *NICOMACHEAN ETHICS*, 335-322 BCE

*"Robbing the world of friendship is like robbing the world of the sun."*
*"Friendship improves happiness and abates misery, by the doubling of our joy and the dividing of our grief."*

—CICERO, *DE AMICITIA (ON FRIENDSHIP)*, 44 BCE

*"Friendship is the source of the greatest pleasures, and without friends even the most agreeable pursuits become tedious."*

—THOMAS AQUINAS, *SUMMA THEOLOGICA*, 1265-74

THE GRECO-ROMAN WORLD, WHICH EXTENDED from the sixth century BCE to the fourth century CE, was grounded in friendship among men. Male citizens spent most of their time in the company of other men, whether at the gymnasium, the marketplace, the Senate, or a private banquet. Their everyday lives were bound up in a system of reciprocity, requiring friends to provide services and even material assistance to one another when needed. The Latin phrase *manus manum lavat*, which is usually translated as "one hand washes the other," is based in the Greco-Roman notion that friends were obliged to support one another and return favors, especially in the political realm. Succinctly put, Greeks and Romans were guided by the principle that one should help friends and harm enemies. In times of war, men slept together in tents and stood side by side in the bitter cold to defend their people against hostile forces. Women, of course, were excluded from all these activities. Whatever friendships they might have had among themselves did not affect the public face of friendship.

Philosophers wrote lengthy treatises about the significance of friendship, and these became standard reading for later generations reared on the classics. Thus the pronouncements of Aristotle and Cicero on friendship have had currency among early Church fathers, medieval clerics, Renaissance humanists, eighteenth-century thinkers, and even a few twenty-first-century intellectuals.[16]

One eminent scholar who has written about friendship in the classical world defines it as "a mutually intimate, loyal and loving bond between two or a few persons" that does not derive primarily from family or tribal connections.[17] Another scholar defines classical friendship as a personal and informal relationship based on reciprocity, choice, and at least the illusion of equality.[18] With these main criteria in mind, Greek men earnestly debated the permutations of friendship at drinking parties called symposia, in lengthy tracts, and in letters, as well as in private conversations.

Ancient Greeks talked about friendship between men with the serious enthusiasm we reserve today for our discussions of relationships between lovers, spouses, or family members.

### Aristotle

Classical philosophers presented friendship not only as an enjoyable end in itself but also as a means to the good life, by which they meant the virtuous life: friendship should help a person become a morally better human being. Aristotle (384–322 BCE) expounded on this idealized vision in his *Nicomachean Ethics*, which identified three kinds of friendships: the first, and lowest, based on utility; the second on pleasure; and the third, or highest, on virtue. In the first and second categories, friends are loved to the extent that they are useful or pleasant to each other: "Such friendships are easily dissolved . . . If the one party is no longer pleasant or useful, the other ceases to love him." By contrast, true friendship, based on a virtuous character, has the potential for enduring over time.

For Aristotle, true friendship can be achieved only between two people bound together by mutual affection, who care more for the well-being of the other than they do for themselves. Ideally, there is such a similarity of heart and mind that best friends become "one soul in two bodies." This concept of friendship, representing the best that Aristotle could imagine, may come across as excessively lofty and ethereal by present-day standards. How many friends today would want the total merger that "one soul in two bodies" implies? And while it is common enough to find two people bound together by affection, how many friends put the well-being of the other party above their own? Such a definition may still represent an ideal in marriage and parenting as well as in friendship.

Even as we quibble with the desirability of some of Aristotle's pronouncements on friendship, they are always worth pondering. Consider the distinction he makes between love (*eros*) and friendship (*philia*). Since eros depends mainly on physical attraction and passionate emotions, it cannot—according to Aristotle—lead to lasting attachments, which is why amorous young people fall in and out of love so quickly. Yet Aristotle's separation of friendship from erotic love is not always hard-and-fast. Given the permissive attitude toward sexual relations between men and boys that was prevalent in ancient Greek society, he is forced to concede that some lovers do become friends, though most of these friendships vanish when the boy loses the bloom of youth.

Same-sex love and friendship in ancient Greece are the subjects of much current scholarship, which tends to remind us that today's gay and lesbian relationships are not identical with those of the past. What we assume to be egalitarian relationships today were anything but egalitarian in ancient Greece. There, same-sex erotic relationships were essentially asymmetrical, with an older male assuming responsibility for the care of a younger one. The lover, says Aristotle's teacher Plato in the *Phaedrus*, experiences passionate desire, or eros, whereas the younger beloved is expected to feel friendly affection, or philia, for the care he receives. Here as elsewhere in our consideration of friendship over the course of 2,500 years, we are called upon to situate it within the framework of a particular culture, whose norms may be very different from our own.

Aristotle was more clear-cut on the number of friends one could or should have, maintaining that "one cannot be a friend to many people in the sense of having friendship of the perfect type with them, just as one cannot be in love with many people at once." These days, the question of just how exclusive one must be in making and keeping friends is open to interpretation. What would Aristotle think of Facebook friends, who communicate highly personal news and thoughts with large numbers of people, often known to one another only through online connections?

It is obvious that in his observations on friendship Aristotle is thinking only of men, with very few exceptions. He grants that both fathers and mothers feel a sort of "friendship" for their children. He allows women and "womanly men" a propensity for sharing grief with companions in sorrow. In contrast, the "manly man" withholds his feelings from his friends so as not to cause them pain. To this day, the philosophy that real men don't cry remains part of the Western tradition.

It comes as a surprise to find among Aristotle's male-centered assertions the statement that "between man and wife friendship seems to exist by nature." He argues that a husband and a wife live together not only for the sake of begetting children but also so that each can contribute his or her strengths to the well-being of the household. The issue of friendship between husbands and wives will become more pressing in later centuries, when women begin to rise toward social equality with men.

What Aristotle wrote concerning friendship between spouses has particular relevance in twenty-first-century America. Often today, even when a wife or husband has friends outside the family circle, they may think of their spouse as their best friend. The man who limits his friendships to colleagues at work and a best friend in his wife may not feel that he needs other friends. This attitude would have astonished Aristotle, given his conviction that men need male friends for personal happiness and that society is held together by male bonding.

Of the various works on friendship written by Greek philosophers, Aristotle's is the most comprehensive and the most accessible, and if we occasionally substitute the words "men and women" for "men," we can create a discussion that includes both genders. Yet even without that sleight of hand, there are formulations in Aristotle's own words that ring true for both men and women today. Take a second look at those listed at the beginning of this chapter, and then add, as additional food for thought, the following: "A friend is a second self."[19]

*Epicurus*

Epicurus, Aristotle's younger contemporary (341–270 BCE), also esteemed friendship (philia) above all other relationships. Though his writings have survived mainly in fragments, the theme of friendship stands out in any reconstruction of his complete philosophy, and friendship is always presented as essential to "the blessedness of the perfect life."[20] Epicurus was somewhat controversial in his own time, and his misconstrued reputation as a pleasure-seeking immoralist has done nothing to add to his posthumous glory. Today we use the word *epicurean* to refer to a person who is devoted to various forms of sensual pleasure, ranging from good wine to sexual enjoyment. This is a total misunderstanding of Epicurus. His philosophy, based on the avoidance of physical and mental pain, downplayed excess of any kind and emphasized, instead, the search for personal happiness, serenity, and the joys of companionship.

Epicurus put his philosophy into practice in the Garden— an open area near the western gateway to ancient Athens— where he met regularly with friends and disciples for the sheer pleasure of conversation. Amazingly, some of these friends and disciples were women. They were probably not "respectable" wives from the citizen class, and they may even have included prostitutes and slaves, but whoever they were, they were welcome to join the parties, where simple food was offered and philosophical discussions constituted the main course. There they would have been treated as men's equals, as they all discussed such heady questions as whether there was an afterlife (Epicurus did not believe in it) and whether one should die for a friend rather than betray him (Epicurus believed so).

Yet while granting friends the capacity for such an altruistic act, Epicurus had no compunctions in admitting that friendships were often based in self-interest and personal need. He went so far as to say that "every friendship is desirable for its own sake but has its beginning in assistance rendered."[21] You can imagine how Aristotle would have reacted to this utilitarian view, which contradicted his own core position that true friendship was based in virtue rather than utility or pleasure. Still, both Aristotle and Epicurus lived at a time when "assistance rendered" was the norm in friendship, at least among Greek men of the citizen class. We know too little about the lives of men from the lower class, and even less about the lives of slaves, to determine whether this system of reciprocity filtered downward, or whether it applied to women.

Aristotle would also have taken issue with Epicurus's view that one should have a considerable number of friends rather than a few good ones. In fact, Epicurus counseled his followers to cultivate as many friends as possible—you never know when you are going to need them—with the caveat that someone "who is continually seeking help" cannot be a true friend.[22]

For Epicurus, philosophy was meant to be a practical guide to life. Today he might have been a psychotherapist, coach, or counselor. If he were speaking to us directly, he would urge: *Look to your friends. They will share your pleasures, comfort you in your sorrows, and help you attain that peace of mind which is the true goal of philosophy and the greatest blessing of life.*

### Cicero

Almost three centuries separate Aristotle and Epicurus from the Roman orator, statesman, and philosopher Cicero (106 to 43 BCE). In his book on friendship, *De Amicitia*, Cicero conjured up the spirit of Gaius Laelius to be his spokesman. Laelius, a celebrated Roman statesman from the prior century, had enjoyed a close friendship with Scipio Africanus, a famous military hero. These two legendary men represented for Cicero the epitome of friendship because the relationship contributed not only to each man's personal happiness but also to the civic good.

In his treatise Cicero imagines a dialogue between Laelius and two other speakers, who are little more than yes-men. Laelius tells them that friendship is the best of life's offerings, superior to health, wealth, power, honor, and pleasure, which are all unstable and dependent on fickle Fortune. He urges them to put friendship ahead of all other human concerns, for there is "nothing so suited to man's nature, nothing that can mean so much to him, whether in good times or in bad." He points to his own long, satisfying friendship with Scipio, concluding that "my life has been rich and good, simply because I spent it in Scipio's company. He and I stood side by side in our concern for affairs of state and for personal matters; we shared a citizen's home and a soldier's tent; we shared the one element indispensable to friendship, a complete agreement in aims, ambitions, and attitudes." This description of citizens and soldiers united in their attitudes and actions represents the highest ideal of friendship among the ancient Romans.

Obviously women did not enter into this scenario. By the time of Cicero, Roman women were certainly freer than the women of Greece had been. Whereas married Greek women had been confined to the female section of their homes and were certainly not present at symposia, the wives of Roman citizens enjoyed both female friends and mixed-sex company. Still, Plutarch (46–120 CE), who had lived in both Greek and Roman societies, reminded us long after the death of Cicero that women's friendships were often constrained by their husbands. In his "Precepts on Marriage," he went so far as to say that "a wife ought not to have friends of her own but ought to share those of her husband."[23] And though we might think of this as ancient history, there are still parts of the world today where a woman marries into her husband's family and is expected to relinquish not only her own family but also the friends of her youth.

But to return to Cicero on friendship: his mouthpiece, Laelius, agrees with Aristotle that virtue is the basis for true and perfect friendship. Laelius gives concrete examples of how virtuous friends should act in certain situations. We must not ask our friends to do anything morally reprehensible, especially in the public sphere, where greed for power and money sometimes puts friendships to the test: "Wrongdoing is not excused [even] if it is committed for the sake of a friend." Virtuous friends—and these are the only kind Cicero respects—must be guided by the premise that we ask of our friends only that which is honorable.

Within the constraints of virtue, friends are allowed full rein over their personal interactions. Friends should "share with each other, without reservation, all their concerns, their plans, their aims." They should assist each other when either party is in need. Friendships strengthen through time and hardship, as expressed in the classical saying that people must eat many a peck of salt together if they are to know the full meaning of friendship.

Laelius/Cicero even gives practical advice about how to end a friendship that has gone wrong. If it is necessary to break with a friend, he prefers that the friendship "should seem to fade away rather than to be stamped out," for fear that it would create hard feelings or turn into serious personal enmity. "There is a degree of respect which we must pay to a friendship," even one that has turned sour.[24]

Cicero, like Epicurus, was certainly aware that friendship in his era was based on a system of reciprocal favors, which seems to contradict the ideal of selfless attachment. In his personal letters, there are times when Cicero himself eloquently employs the language of friendship as a prelude to asking for help.[25] Nonetheless, Cicero clung to his vision of idealistic friendship and passed it on to those who have continued to read *De Amicitia* for more than two millennia. It was highly influential during the Middle Ages, when theologians tried to integrate the Ciceronian view of friendship into a Christian context. It continued to be read during the Renaissance, when the education of young men often included learning passages of *De Amicitia* by heart. Even when Christian values overtook those of antiquity and friendship became subordinate to one's relationship with God, Laelius and Scipio Africanus—as well as the earlier Greek literary examples of Achilles and Patroclus, and Orestes and Pylades—could be interpreted as prototypes for the close male companionship enjoyed by Jesus and his disciples.[26]

*Saint Augustine*

One of the earliest Christians to write about friendship was Augustine of Hippo (354–430). Born in North Africa, he was educated at the University of Carthage, embraced a non-Christian sect called Manichaeism, and lived what he later considered a life of error and sin, until he relocated to Milan and converted to Christianity in 386. In his autobiographical *Confessions*, written around 394, Saint Augustine recounts the acute pain he experienced at the death of a beloved friend. In the words of Augustine, addressing God: "You took him from this life after barely a year's friendship, a friendship sweeter to me than any sweetness I had known in all my life." After experiencing the joys of friendship—talking, laughing, joking, and reading together—Augustine was devastated by his irreplaceable loss.

His friend's death caused him to grieve in a manner that we associate more readily with lovers or spouses when they have lost a lifelong partner. In modern parlance, he became depressed: "My eyes sought him everywhere, but he was missing; I hated all things because they held him not." Realizing too late that his friend's soul and his own had become "one soul in two bodies" (echoing the words of Aristotle), Augustine found it unbearable to exist "only half alive."[27]

What differentiates Augustine's conception of friendship from that of his Greek and Roman predecessors was the Christian cloak he threw over it. For example, he came to appreciate the baptism that his friend had unknowingly received when he was near death's door. For a short period after being baptized, his friend was restored and imbued with a Christian conscience, so that he rebuked Augustine— then an unbeliever—for making fun of baptism. Much later, when Augustine himself had been baptized, he came to believe that friendship, no matter how deep, is perishable if it is not bound up in the love of God. Augustine ascribed to God what Greek and Roman thinkers ascribed to virtue. Just as virtue was, in classical philosophy, the necessary element uniting two friends, so the love of God was for Augustine the all-encompassing light infusing true friendship. This primary focus on God as taking priority over any flesh-and-blood relationship is the hallmark of Christianity.

### Saint Thomas Aquinas

During the Middle Ages, the great Catholic theologian and philosopher Thomas Aquinas (1225–1274) gave friendship a place of supreme prominence in his religious doctrine. Looking back to Augustine, and even more to Aristotle, he began the section on love in his monumental *Summa Theologica* by asking whether the Christian virtue of charity can be understood as friendship. Starting with objections to this assumption, which is Aquinas's usual method of argumentation, he comes eventually to his core thesis, found in the parting words of Jesus to his disciples: "I call you not servants . . . but I have called you friends" (John 15:15). If Jesus thought of his beloved associates as friends, then Christians wishing to emulate their Lord should act similarly and place friends above all other human relationships.

Again following Aristotle, Aquinas distinguishes between friendship and love. He argues that the latter is based on desiring something for oneself, whereas the former concerns the good that one desires for another. Though both friendship and erotic love may be seen as forms of love, only friendship has the quality of mutual benevolence that makes it, according to Aquinas, superior to erotic love.

Like his spiritual model, Saint Augustine, Aquinas insists that the love of God is the gateway to loving others; only through the love of God can one internalize the spirit necessary for true friendship. To "build up a friendship" with God and then with other human beings sums up, according to one twenty-first-century Catholic Church spokesman, the entire *Summa Theologica*.[28]

For anyone not schooled in philosophical argumentation and theological debate, Saint Thomas Aquinas is a difficult, dry, scholastic read. He has none of the personable style of Saint Augustine, nor the down-to-earth practicality of Cicero. Yet occasionally he pens a straightforward sentence that reveals his human side and endears him to a contemporary reader. "Friendship," he writes, "is the source of the greatest pleasures, and without friends even the most agreeable pursuits become tedious."

### Monastic Friendships: The Example of Saint Anselm

Though most medieval clerics lived in urban or rural communities, preaching to the faithful and interacting with parishioners of both sexes, some chose to isolate themselves from the secular world and live in monasteries. Already in the third and fourth centuries Christian monasteries and convents were being founded in Egypt, followed by similar establishments in Europe, where they would become powerful institutions. From the sixth century onward, the *Rule of Saint Benedict*, written by Benedict of Nursia (circa 480–547), established the regulations for most European monastic men and women. With his own abbey, Monte Cassino in mind, Benedict wrote the guidelines for an orderly cloister where all members lived, ate, slept, prayed, and worked together. Given the primary emphasis on communal activity, individual or "particular" friendships were discouraged because they were considered disruptive to the general good. Benedict's second rule specifically warned the abbot against loving "one more than another."

Unlike classical philosophers, especially the ancient Greeks, Christian thinkers condemned sexual love between men. Yet this did not prevent some Christian men from feeling attraction to one another and expressing themselves like lovers. Read the words of Anselm (1033/34–1109) writing to two of his kinsmen, who had arrived to see him at Bec Abbey in Normandy, after he had already left on a trip to England: "When I heard that you had come such a long distance to seek my face I cannot express what great joy flooded my heart . . . My eyes desire, most dearly beloved, to see your faces, already my arms stretch out to take you in their embrace. My mouth yearns for your kisses."[29] Or consider these words, from his letter to the monk Maurice: "Although the more I love you the more I want to have you with me, yet because I cannot have you I love you even more."[30] And again to his fellow monk Gundulf, after Gundulf left the monastery at Bec where they had been together: "Wherever you go my love follows you; and wherever I may be, my longing for you embraces you."[31]

These are hardly the words we would expect from an ascetic monk. Taken out of context, they seem to convey homoerotic desire; yet as one situates them within Saint Anselm's entire correspondence, one sees a pattern of affection that is always subordinate to Anselm's concern for the other person's well-being and for the entire monastic community of which they were a part. Brian McGuire, author of a hefty tome on medieval friendships among monks, puts it this way: "He looked at friendship as a way to enrich the content of monastic life, which Anselm considered the best and often the only way to reach paradise."[32] McGuire traces the language of love within these letters to a "rebirth of friendship" in the last half of the eleventh century that originated in certain French and German cathedral schools, most notably in Chartres, Reims, Beauvais, Worms, and Hildesheim.[33] Based on the study of Latin grammar and such classical writers as Aristotle and Cicero, the new curriculum in these schools honored friendship as a noble and natural enterprise that could and should be incorporated into Christian life. Both secular and religious students practiced letter writing according to manuals that taught them the proper rhetoric for expressing friendship, be it a merely formulaic greeting or a passionate declaration of love. Collections of letters from this period written by students and their masters, as well as by various ecclesiastical figures, have been stored away for a thousand years in German and French archives.

We should not forget that all these letters were destined to be read aloud, first by the recipient and then by many others who had access to copies. Saint Anselm wrote to his friends in a semipublic mode, both to express his personal feelings toward individual men and to display an idealized picture of brotherly love.

Anselm encouraged personal friendships among the monks in his charge when he was prior and then abbot of the monastery at Bec. Later, when he became archbishop of Canterbury in 1093, he extended that vision to the entire clergy. (After 1066, when William the Conqueror invaded England, clergymen from France held high posts across the Channel; the Latin language united them with their English brethren.) Anselm's letters are testimonies to a renewed celebration of friendship within monastic communities. Whereas earlier ascetic groups had warned against the dangers of friendship, Anselm embraced it, as long as one's personal love for another did not stand in the way of the love one felt for all the other brothers in the community. Though he waxed eloquent about the value of personal feelings in a new way and in a new language, he always subordinated such feelings to the overall goal of loving God and all men.

Today Saint Anselm is not as famous as his spiritual descendant, Saint Francis of Assisi. Saint Francis comes to mind more readily when we think of love, especially because of his love for all living creatures. The city of San Francisco, which bears his name, is often called the city of love; it contains a wide diversity of peoples living together with a remarkable degree of tolerance and harmony. Twenty miles away from San Francisco, the town of San Anselmo can also claim to be a community of love in honor of its patron saint, who significantly altered the map of friendship in Church history.

Because of Anselm and others who followed in his footsteps—for example, Bernard of Clairvaux, Aelred of Rievaulx, John of Salisbury, and Peter of Blois, who all lived in the twelfth century—personal friendships between men of the cloth were granted a new stamp of approval. Because clergymen could not have wives or mistresses (though many were known to have had concubines, commonly referred to as "priests' wives"), church mores allowed them to express their affection for one another with remarkable openness. The letters they left behind trace a network of friends ranging from popes, archbishops, bishops, cardinals, abbots, and priors down the hierarchy to parish priests and monks. Whether they wrote to seek Church influence, argue a theological point, or convey personal news, they looked to one another for emotional and intellectual support. Friendship became, once again, a highly prized commodity, as it had been among the disciples of Jesus. To emulate the earliest Christians, medieval monks assumed the mantle of loving brothers when they worked, prayed, and ate together. Whatever differences and rivalries they experienced among themselves were always subordinated to their primary goal, which, in the words of Saint

Augustine, was "to live harmoniously in the house and have one heart and one soul seeking God."[34] Nuns, too, as we shall see in the next chapter, relied on friendships with members of their own sex to form a community of Christian women dedicated to the service of God.

# PART TWO

When Women's Friendships
Entered into History

# THREE

# PREMODERN NUNS

*"I bore a deep love for a certain noble young woman . . . She joined herself to me in loving friendship, and comforted me in all my trials."*

—HILDEGARD OF BINGEN, *THE LIFE OF HOLY HILDEGARD*, C. 1170-80

*"Let no Sister embrace another or touch her on the face or hands. The Sisters should not have particular friendships, but should include all in their love for one another, as Christ commanded His disciples."*

—TERESA OF ÁVILA, CONSTITUTIONS OF THE DISCALCED CARMELITE ORDER, 1581

THROUGHOUT HISTORY, MOST GIRLS AND women have lived within families, first with their parents and siblings and then with their husbands and children. In polygamous societies, children have often lived in the company of plural mothers, and wives with co-wives. Women's friendships were, and often still are, formed primarily with their kin—sisters, cousins, aunts, and sisters-in-law—rather than with people outside the family circle.

Another option became possible for women when some early Christian communities set aside a separate part of the monastery for nuns. In these sex-segregated enclaves, nuns, like their male counterparts, withdrew from society and dedicated themselves to chastity, poverty, and obedience. Here there was no question of starting new families, since nuns and monks took vows to renounce sexual activity.

As one historian of Catholic nuns has suggested, segregation by sex had the merit of "introducing women to the great joys and rewards of friendship with one another."[35] After taking their vows, nuns counted almost exclusively on their sister nuns for the ties of friendship they would have made in the secular world.

Monasteries with separate sections for women, as well as women-only convents, were already sprinkled throughout Egypt and parts of Europe by the fourth century. Saint Augustine encouraged the nuns in a convent that had previously been headed by his sister to love one another, but he was wary of friendships that might lead to sexual intimacy. Thus he instructed nuns to go to the public baths only in groups of three or more, and whenever a sister had to leave the convent for any reason, she was instructed to go with someone chosen by the prioress rather than by herself. This early fear of homoerotic friendships subsequently echoed throughout the history of Christian monasteries and convents for centuries to come.[36]

When Saint Benedict founded the monastery of Monte Cassino in the sixth century, his sister, Scholastica, established a convent nearby with a group of female disciples. During the following centuries, Benedictine monasteries for men with sister communities for women spread from Italy into France, Germany, and England. Other religious orders founded in the twelfth and thirteenth centuries—Cistercians under Saint Bernard, Franciscans and the Poor Clares under Saint Francis, and Dominicans under Saint Dominic—also created separate dwellings for nuns.

Many of these nunneries accepted only upper-class women, with the proviso that they bring substantial "dowries" with them. Such convents catering to the well-born had definite snob appeal; families knew that their daughters would associate only with others of their class and perform tasks suitable to their station, such as reading, writing, singing, and embroidery. Domestic service and manual work could be left to either unpaid servant nuns or paid lay servants from humble backgrounds. The servant nuns did not necessarily profess vows, and they tended to be confined to the kitchen and service areas. Only the veiled, or "choir," nuns from wealthy noble families were involved with convent administration and could rise to the positions of treasurer, teacher of the novices, prioress, and abbess. In this way, the secular class system was replicated within nunneries.[37]

What do we know about female friendships in these establishments? Since medieval nuns were expected to read and recite the Latin liturgy and some could also write in both Latin and the vernacular, they have left behind a considerable body of letters, memoirs, and theological treatises. This level of female literacy was highly exceptional throughout the

Middle Ages; indeed, such was the fear of educated women that French and Italian advice manuals authored by men during the thirteenth and fourteenth centuries specifically stated that women should not learn to read or write "unless they are going to be nuns."[38] In addition to the writings left behind by nuns, there are extant reports of cloistered women's lives written by men—local priests who conducted Mass and heard the women's confessions; bishops and archbishops who made annual visitations; and other church authorities charged with overseeing the women's communities. From these documents we can tease out some strands from the intricate knots that tied convent women to one another.

### Hildegard of Bingen

The brilliant, creative, passionate Hildegard of Bingen left us tantalizing remnants of her and her sister nuns' time and place. Born around 1098, Hildegard was the tenth child of an upper-class, though untitled, family from the Rheinhessen region of what is today Germany. At the age of eight, she was given as a tithe by her family to the care of Jutta of Sponheim, an aristocratic adolescent who had already expressed her desire for a religious vocation. In 1112 Jutta took the vows of an anchorite, intending to live in solitary confinement among the monks at the Benedictine monastery of Disibodenberg, and Hildegard, age fourteen, followed her there the same year.[39]

The practice of enclosure meant that neither Jutta nor Hildegard was permitted to leave her cell, possibly with the exception of attending chapel. They probably received their food and other necessities through a window or door with a grille. The account of Jutta's life, *Vita Jutta*, written by the monk Volmar with Hildegard's help, does not provide much material detail. What is certain is that Hildegard and Jutta depended on each other and developed a friendship under extreme circumstances.

Jutta, the elder of the two, instructed Hildegard in the basics of religious life and the singing of the psalms. Together they followed the Benedictine daily schedule of prayer, work, study, meals, and sleep. From Jutta, Hildegard learned sufficient Latin to read and write, though she later bemoaned her lack of instruction beyond the basics. They helped each other with their needlework, intended for the church. Certainly they must have counted on each other for the care of bodily functions, such as their monthly menses and the daily disposal of waste products. Of course, none of the written documents mentions these unmentionables.

What they do mention are Hildegard's frequent illnesses and the visions she began to experience early in life. These visions, seen at that time as the possible mark of a God-chosen mystic, would eventually appear in her remarkable writings and music, but as a girl and young woman she kept them to herself: one could never know if visions were the work of God or the devil. She could not, however, hide the pain they caused her—the kind we associate today with migraine headaches. Jutta, we imagine, would have been at her bedside during these crises, perhaps covering Hildegard's eyes and praying for her. Hildegard, when she was up and about, served her teacher and friend as an assistant, increasingly so during the years when Jutta's renown as a holy woman drew to the monastery other noblewomen who wanted to share her religious life. In time, Jutta established around herself and Hildegard a small female community, and Jutta even accepted the position of *magistra*, head of the nunnery within the monastery.

Jutta died in 1136 at the age of forty-five. She had been enclosed within Disibodenberg for twenty-four years. Hildegard and two other women prepared her for burial according to the age-old rituals, which included washing the body and arranging the limbs. Jutta would be buried on the grounds of the monastery and later reburied in front of the chapel altar.

That same year Hildegard was appointed to succeed her mentor as magistra, a position that would grant her considerable authority, though she remained subservient to Abbot Cuno. She began her tenure by writing religious music, and songs for the nuns to sing. Some of this music has survived and is still performed today by medieval-music groups. (Listen, for example, to the stirring all-female version of Hildegard's "O Jerusalem," sung by the Deller Consort.) We have to imagine her instructing the nuns in the intricacies of notation and voice as they prepared to sing at the daily liturgical service, known as the Divine Office. And it is not hard to imagine the sense of fellowship that would have emerged among the nuns from singing together.

Within such an atmosphere, the eighteen women in residence at Disibodenberg in 1140 had much in common. All had come from noble families, bringing dowries in their role as brides of Christ. All had chosen, or were forced by their families, to sequester themselves from the secular world and to fulfill the Benedictine vows of chastity, poverty, and obedience. But as for the vow of poverty, it seems they were not all inclined to abandon their lavish dress and long hair, which was permitted for the noblewomen at Disibodenberg but not for women of lower classes at other nunneries, who were obligated to have their heads shorn several times a year. Hildegard approved of the unbound hair with golden crowns and ankle-length white garments worn by her virginal nuns when they sang the psalms on feast days. Such scenarios originated in her inner visions, which she had begun to write down with the help of the monk Volmar—her spiritual adviser and loving friend for more than thirty years. News of Hildegard's visions and theological writings was soon transmitted from Abbot Cuno to his superior, the archbishop of Mainz, who sent a delegation to investigate. Was this Hildegard a bona fide seer or a fraud? The delegation, satisfied that she was the real thing, sent its report to the pope, and he, too, became convinced that she was a genuine God-chosen mystic.

Around 1150, with backing from the pope and her own abbot's approval, Hildegard convinced the authorities that she should found a new establishment exclusively for nuns. That Hildegard managed to create a new cloister out of monastic ruins situated on a distant hill called Rupertsberg is only one of her many amazing accomplishments. There she continued to direct the lives of her female charges, including her favorite, Richardis of Stade. This example of a "particular" friendship is well documented in letters written by Hildegard and the responses of her correspondents. It demonstrates how intensely one nun could feel for another nun, despite the opposition to personal friendships stated explicitly in *The Rule of Saint Benedict*.

## Hildegard and Richardis

Richardis of Stade, the daughter of a noble family related to Jutta of Sponheim, was more than twenty years younger than Hildegard. She not only received religious instruction from her magistra but also helped in the preparation of Hildegard's first manuscript, *Scivias*. This is how Hildegard described her: "For while I was writing the book *Scivias*, I bore a deep love for a certain noble young woman . . . She joined herself to me in loving friendship in everything, and comforted me in all my trials, until at length I finished that book."[40]

Hildegard's "loving friendship" with Richardis was problematic because she was much younger than Hildegard, and inferior in rank. But as we shall see here and elsewhere, equality of age and station are not always essential in friendship, regardless of Aristotle's opinion. Hildegard referred to Richardis as her "daughter" and refused to give up the role of a controlling "mother," even when Richardis was elected abbess of Bassum, another Benedictine monastery. Hildegard could not bear for her beloved friend to be torn away from her. She sent off a series of heartrending missiles—to Richardis's mother, to archbishops, and even to the pope—in her efforts to retain Richardis at Rupertsberg. She described her plight in a letter to Richardis's brother, the archbishop of Bremen: "Now hear me, cast down as I am, miserably weeping at your feet. My spirit is exceedingly sad because a certain horrible man has trampled underfoot my desire . . . and has rashly dragged our beloved daughter Richardis out of our cloister."[41]

And what did Richardis think of all this? We do not know. But she seems to have moved to her new cloister by choice, and she must have been troubled when she received, sometime around 1151 or 1152, Hildegard's doleful letter containing the following passages:

> Daughter, listen to me, your mother, speaking to you in the spirit: my grief flies up to heaven. My sorrow is destroying the great confidence and consolation that I once had in mankind . . .
>
> Now again I say: Woe is to me, mother, woe is to me, daughter. Why have you forsaken me? . . . I so loved the nobility of your character, your wisdom, your chastity, your spirit, and indeed every aspect of your life . . .
>
> Now, let all who have grief like mine mourn with me, all who, in the love of God, have had such great love in their hearts and minds for a person—as I had for you—but who was snatched away from them in an instant, as you were from me.[42]

Certainly the language is that of an aggrieved friend, perhaps even that of a distraught lover. Over and over again in our study of female friends, we have found examples of women loving other women passionately and even physically, with caresses and kisses, though not necessarily sexually. These are what we call "loving friendships." Though Hildegard clearly wanted Richardis to remain within the confines of Rupertsberg, it was not to be. Richardis spent two years at the Bassum monastery before she succumbed to illness and died. Richardis's brother, the archbishop of Bremen, had the sad and difficult task of informing Hildegard of his sister's death. He was remorseful enough to give Hildegard the satisfaction of knowing that during her last rites Richardis had tearfully expressed her desire to return to Rupertsberg, at least for a visit. And he added, humbly, "Thus I ask as earnestly as I can, if I have any right to ask, that you love her as much as she loved you . . . May God, who repays all good deeds, recompense you fully in this world and in the future for all the good things you did for her, you alone, more even than relatives or friends."[43]

This moving story portrays the depths of feeling that could exist between one nun and another, even when there were substantial differences in age and status. It demonstrates that intense friendships could develop within austere cloisters, even though the nuns were warned against them: there was always the fear that such attachments could dilute one's commitment to God. In addition, by showing preference for each other two friends could undermine the sense of cohesiveness essential to a harmonious cloister. Even worse, highly emotional friendships could lead to sins of the flesh. From its earliest days, the Church inveighed against homoerotic relations. For example, in addition to the concerns expressed by Saint Augustine and Saint Benedict, Caesarius of Arles (470–542) explicitly counseled the fledgling nuns in his community to avoid forming "unsuitable intimacy" with others. "No one," he stated, "shall have a secret intimacy or companionship of any kind." Better to maintain an aura of distance around oneself than to fall into the carnal temptations provoked by an emotional attachment.[44]

Most nuns were limited to friends within the cloister. But occasionally they were able to form friendships with visiting family members and other acquaintances, though they could meet only on opposite sides of a grille, like high-security prisoners today. And some nuns, through the medium of letters, formed friendships with women in other convents.

### Hildegard and Elisabeth of Schönau

Such was the case of Hildegard of Bingen and her younger peer, Elisabeth of Schönau (1129-65), magistra of the women's section of the double monastery of Schönau. The letters they exchanged from one Benedictine convent to the other included descriptions of mystical experiences, doctrinal discussions, expressions of sympathy, and various concerns issuing from their similar positions.

Like Hildegard, Elisabeth had been enrolled as a pupil in the monastery when she was very young, and she would remain there for the rest of her life. Unlike Hildegard, whose visions had begun in childhood, Elisabeth's dated from when she was already a grown woman. Around 1152, at the age of twenty-three, she began to have the ecstatic trances she described in her letters to Hildegard. During those searing moments, her sister nuns looked after her with loving care, and afterward, some helped her write down her visions. Before long, Elisabeth's brother Ekbert began to circulate them, and she became famous.[45]

Elisabeth was already experiencing the mixed effects of celebrity when she first wrote to Hildegard asking for advice. She complained that false letters had appeared under her name and that her reputation was being sullied by gossip and rumors. Elisabeth assumed that Hildegard, a fellow mystic, would understand her predicament. Hildegard responded sympathetically, with motherly advice. If one had been singled out by God, as they believed they had been, one must remain humble and trust in the Lord. Hildegard's faith was unswerving, even though she, too, was no stranger to inner turmoil, which she expressed openly to Elisabeth: "I too cower in the puniness of my mind, and am greatly wearied by anxiety and fear. Yet from time to time I resound a little, like the dim sound of a trumpet from the Living Light. May God help me, therefore to remain in His service."[46] The two women saw each other as confidantes with whom they could entrust their personal anxieties. After all, what are friends for if not to vent worries without fear of negative consequences?

Convent life encouraged what we would today call mentoring. Older nuns were expected to look out for younger nuns, to guide them as they adjusted to enclosure and struggled to become worthy brides of Christ. Just as Jutta had instructed Hildegard, so Hildegard guided Richardis, and even Elisabeth from a distance. Destined to live together, away from their families and secular society, the younger nuns counted on the older ones to teach them the religious practices necessary for communal life, and ultimately for salvation. The senior nuns did not take these responsibilities lightly. In time, when their situations became more egalitarian through the younger nuns' integration into convent life, the older and younger nuns could become bona fide friends.

Even when a nun left the convent of her youth for another convent, she might still remain attached to her former superior and look to her for advice. Such was the case of Adelheid, who, like her relative Richardis von Stade, encountered Hildegard's opposition when she wanted to accept the position of abbess at a different cloister. From her new post she wrote to Hildegard, begging her to renew the ties of their "earlier friendship" and to pray for her and her flock. Her language is tender and intimate: "You ought to keep me frequently in mind, since, as is well known, I am joined to you in intimate closeness of love and devotion. I do not want the flower, nursed so gently in former days, to dry up in your heart, the blossom that once vitally flourished between the two of us at the time when you were gently educating me."[47] Hildegard's response to Adelheid's emotionally charged epistle has not been preserved, but a later letter suggests that Adelheid and Hildegard corresponded over a period of twenty years, during which time the older woman continued to advise and succor the younger one.

Other abbesses, prioresses, and groups of nuns, whether they knew Hildegard personally or not, wrote to her for her blessing and counsel. An abbess in severe distress insisted that she felt bound to Hildegard "by a deep feeling of love."[48] A worried prioress wondered what she should do about her unruly charges, since, as she put it, "it is my duty to correct the waywardness of my sisters—and this despite the fact that I can scarce fight off the dangers that surround me on every side." Hildegard counseled the prioress to continue to "labor among the daughters of God" as long as she had the ability. She always advised nuns in positions of authority to persevere rather than abandon their administrative obligations.[49]

The abbess Sophia of Kitzingen wrote to Hildegard to announce a forthcoming visit in the company of another nun. Here is how she described her companion: "I am bringing with me a well-born peer of mine, a praiseworthy nun, a sister acceptable in every way, whom the heavenly Father has created as my spiritual sister. It is God's will that the two of us make your acquaintance."[50] There is something so human—so timeless—in the pride the abbess felt for her stellar friend and in her desire to parade this friendship before Hildegard's approving eyes. The abbess Sophia undoubtedly felt that it enhanced her own value to be associated with such an admirable person. But lest we mock this attempt at self-validation through association, how many of us today can honestly say we have never dropped the name of a prestigious friend for a similar reason?

From these letters to Hildegard and others, it is clear that nuns looked to one another for support, affection, and friendship. Like Saint Anselm, whose letters were already in circulation during their lifetimes, they did not shy away from such words as *love, heart,* and *devotion* to describe their feelings. And it is also clear that older, more established nuns, like Hildegard, were treated with a heady mixture of love and respect. It was incumbent upon famous holy women to offer counsel and consolation not only to the nuns directly in their charge but also to those at a distance who sought spiritual mentorship.

### Mechtilde of Hackeborn and Gertrude the Great

A hundred years after the flowering of Rupertsberg under Hildegard, another German cloister became famous for its saintly nuns and their erudition. During its golden age in the thirteenth century, the Benedictine monastery of St. Mary at Helfta in northern Saxony produced the visionary Mechtilde of Hackeborn (c. 1240-98), who was the friend and mentor of another mystic, Gertrude the Great. Through their writings it is possible to glimpse their close friendship, with the older nun instructing the younger one and both of them inspiring their companions.

Mechtilde of Hackeborn entered the cloistered world when she was only seven. Her mother had brought her to visit her sixteen-year-old sister, Gertrude of Hackeborn, at the monastery situated near their home. (Gertrude of Hackeborn is not to be confused with Gertrude the Great, as many have done over the centuries.) For the next ten years Mechtilde was raised in the company of other children under the supervision of her sister, who became the monastery abbess in 1251.

When the community moved from Rodersdorf to Helfta in 1258, Mechtilde went with it.

Mechtilde taught children at the Helfta convent school, as well as music to the nuns. She became known for her outstanding singing voice and her direction of the choir, both of which earned her the unofficial title of *domna cantrix* (lady chantress). Mechtilde's younger contemporary, Gertrude the Great, called her the "nightingale of Christ."[51]

For almost forty years Mechtilde shared the company of the two Gertrudes, assisting her abbess sister in convent administration and serving as the spiritual adviser of her younger friend. Mechtilde and Gertrude the Great chanted the liturgy with the other nuns, and read Scripture together in private. These two choir nuns worked together on their daily household tasks, which included spinning and embroidering. They took care of the sick on a regular basis and assisted the doctors in bloodletting.

In 1291, when she was more than fifty years old and in poor health, Mechtilde of Hackeborn confided her revelations to Gertrude, and eventually allowed her and another nun to record them. Without the help of her sister nuns, Mechtilde's revelations would never have found their way into her *Book of Special Grace*. In this and many other cases, nuns were responsible for documenting the pronouncements of their visionary companions.[52]

Three hundred years after Mechtilde's visions were penned on parchment, those of the Carmelite Maria Maddalena de Pazzi (1566–1607) were also recorded in a collective writing process. Her friends in the cloister transcribed the words she uttered while she was in an ecstatic trance. Then they compared their texts and eventually, after Maria Maddalena's death, edited them for publication.[53] This collaboration in reconstructing a woman's visions can be seen as a very early precursor to the collective work that feminist scholars undertook in the late twentieth century to restore women's experiences to history.

Similarly, the life history of Isabel de Jesús (1584–1648) would not be known to posterity if a female friend had not written it down, since Isabel herself was illiterate. Born into a Spanish peasant family, she spent her childhood years tending flocks and experiencing mystical visions. Later she married an older man; after he left her widowed in her early adulthood, she retired to a convent as a servant nun. There she dictated her life story and visions to a learned sister, Inés del Santissimo Sacramento, who came from a wealthy, upper-class family. The two provided mutual assistance to each other: one took on the role of educator and transcriber; the other cared for her protector in sickness. Later, when Inés became abbess of the convent, Isabel was upgraded to the status of a choir nun. For once the lines of social class were crossed, to the benefit of two nuns from vastly different social backgrounds.[54]

Does this mean that all, or even most, nuns behaved like good friends, without experiencing animosity or dissonance? By no means. Regardless of time or place, it is not uncommon to find incidents of discord within the monastic community. Records of fights between nuns that included physical violence and even murder attempts suggest some of the pit-falls hidden behind an aura of sanctity.[55] Whenever a group of people—however spiritual—lives together, there are bound to be expressions of dissent, rivalry, favoritism, and jealousy, and numerous outbursts that disrupt the communal peace.

### *"Immodest Acts"*

Among the many charges leveled against medieval convents were that they allowed for sexual affairs between the nuns and visiting priests. The occasional birth of a baby gave substance to these charges, as in the case of a French nun at the priory of Saint-Aubin, who bore three children, at least one of whom was known to have been fathered by a priest. Archbishop Eudes Rigaud recorded these events in his register from 1248 to 1269, and he fulminated against the nuns in this convent who lied to cover up for one another.[56] Since we do not have the nuns' version of these happenings, we can only wonder why and how they banded together to help their sister nun through her pregnancies and births. They seem to have felt loyalty to one of their own, even in her disgrace. Perhaps some of them unconsciously identified with this sexually active woman and envied or empathized with her experience. Perhaps the births brought out their own maternal feelings as they coddled the newborn babes. Even knowing that sex for a nun or monk was considered a venial sin, they seem to have stood up to the bishop in an act of female solidarity, which enraged him.

By the time of the Renaissance, some convents were no-torious for their loose morals. Historian Judith Brown was so familiar with documents recording the heterosexual affairs of nuns in early modern Italy that she called them "commonplace," but she was surprised to find that an extensive ecclesiastical investigation that took place from 1619 to 1623 contained a description of the sexual relations between a convent abbess and a younger nun.[57]

Benedetta Carlini, born to modest parentage in the grand duchy of Tuscany, was placed with a group of religious women in the town of Pescia when she was only nine. At the age of thirty she was elected by the other nuns to be abbess of the group, by then a bona fide Theatine convent. She had already begun to have the visions of Christ that would ultimately lead to both her fame and her downfall. Like other female mystics before her, she would fall into a trance and experience great bodily pain when the visions were upon her. Not content with acquiring the reputation of a visionary among the other nuns and even the townspeople, Benedetta staged an elaborate ceremony during which she and Jesus were publicly wed within the convent chapel in a mystical marriage recalling that of Saint Catherine of Siena several centuries earlier. This proved too much for the ecclesiastical officials, who decided to investigate.

What they discovered proved to be beyond their worst fears: not only were Benedetta's visions, as she recounted them, filled with dubious claims, but she had also enticed a younger nun, Sister Bartolomea Crivelli, to come to her bed and engage in "immodest acts." According to the testimonial of Sister Bartolomea, the abbess had kissed her "as if she were a man" and spoken to her words of love. "And she would stir on top of her so much that both of them corrupted themselves. And thus by force she held her sometimes one, sometimes two, and sometimes three hours."[58] For about two years, Benedetta and Bartolomea were, by their own admission, lovers.

We have included this story not because it is typical (as some eighteenth-century scandalmongers would have us believe) but because it represents a case in which friendship crossed over into the territory of sexual relations. Some female friendships throughout the ages have veered in this direction, especially where women were segregated, by choice or force, within same-sex communities.

The clerics who questioned these two women had no understanding whatsoever of the feelings the two nuns had for each other. Although hundreds of cases concerning sex between priests and nuns, as well as homosexual behavior between men of the cloth, had come before the ecclesiastical courts, they had never before dealt with the mechanics of sexual gratification between women. In the end they judged the relations between Benedetta and Bartolomea to be the work of the devil, with which both women concurred. Far better to ascribe their "immodest acts" to demonic possession than to their own free will, which could well have incurred capital punishment. In the end Bartolomea got off lightly, whereas Benedetta Carlini was isolated from the other nuns in her convent for the last thirty-five years of her life.

Such total isolation was rare. Opportunities for striking up a meaningful friendship varied greatly among European convents, depending on the country, time period, ecclesiastical authorities, and personality of the abbess or prioress in charge. In thirteenth-century Spain, for example, nuns were allowed to play chess, despite Church prohibitions against the game. We know this from King Alfonso X's magnificent 1283 *Book of Games*, which contains an image of an older nun on one side of the board instructing a novice opposite her in the rules of the game.

Some nuns specialized in medical assistance. They participated in teams that took them into the outside community; for example, some worked in the Parisian hospital Hôtel-Dieu. A fifteenth-century manuscript titled *Livre de la vie active des religieuses de l'Hôtel-Dieu* ("The book of the active life of the nuns from the Hôtel-Dieu") describes their activities and contains a detailed image of four black-robed nuns ministering to bedridden patients while younger nuns, half their size, look on and learn. Once again, we ask ourselves what friendships were made among these women as they went from their dormitories to the beds of the sick. Working together and sleeping in the same room may have provided opportunities for nuns to discuss not only their shared concerns (a sick patient, the attending doctors) but also their personal stories.

### Saint Teresa of Ávila

The personal story of Teresa of Ávila (1515–82), as she recounted it in *The Book of Her Life*, indicates that she had an unusual gift for friendship. In fact, she was so friendly and so talkative that, for a long period of her life as a nun, she was conflicted over her ties to her friends and her ties to God. In her own words: "For more than eighteen of the twenty-eight years since I began prayer, I suffered this battle and conflict between friendship with God and friendship with the world."[59]

Teresa spent her first two decades as a nun at the Carmelite Monastery of the Incarnation in Ávila. Then, in 1562, after she had a vision that caused her to want to perfect the Carmelite Rule, she and several of her friends established a reformed branch of her order, called the Discalced (barefoot) Carmelites, in the Convent of San José in Ávila. There they instituted strict enclosure and limited themselves to infrequent contacts with outside relatives and acquaintances, in order to focus more fully on prayer and their inner lives.

In the same year she composed *The Book of Her Life* at the behest of her confessor, so as to explain her visions and her practice of "mental prayer" to a larger audience. The book did indeed reach beyond her small community and was read by priests, theologians, laypersons, and even the bishop of Ávila. If her visions had taken her into the future, she would have seen that her book is still being read in the twenty-first century, not only in Ávila but also across the globe.

Teresa wrote a second book, *The Interior Castle*, to clarify certain aspects of mental prayer for other Carmelite nuns, since, as she put it, "women best understand each other's language."[60] She herself certainly knew how to talk to other women, clearly and sympathetically. From her own account Teresa was a consummate conversationalist, by nature an extrovert, given to friendship from the time she was a child.

Throughout *The Book of Her Life* Teresa refers to numerous friends, named and unnamed, who provided sustenance when she direly needed it. There was the older nun from the Incarnation, who accompanied her to a place famous for its cures when she was severely ill. In Teresa's words: "I was brought there with much solicitude for my comfort by my father and sister, and my friend, who had come with me, for she loved me very dearly."[61]

Then there was the "widow of high nobility," known to be Doña Guiomar de Ulloa, who housed Teresa for many days when she was going through a crisis at the monastery caused by the departure of her first Jesuit confessor. Doña Guiomar had inherited a small fortune from her husband, which she used for charity and to help found Teresa's new Convent of San José. With Teresa's encouragement the wealthy widow also tried out the Carmelite life at San José, but she was not sufficiently strong to endure it.

Teresa also became friends with one of her confessors, who ended up confessing more to her than she to him. In Teresa's words: "By reason of the strong love he had for me, he began to explain to me about his bad moral state. This was no small matter, because for about seven years he had been living in a dangerous state on account of his affection and dealings with a woman in that same place; and, despite this, he was saying Mass." Teresa admitted that she, too, "loved him deeply" and was afraid that their frequent conversations were causing him to love her even more. Looking back upon their interactions, Teresa argued, a bit like Cicero, that one should not persevere in a friendship if it causes one to do wrong. In her words: "This is the kind of nonsense that goes on in the world . . . that we consider it a virtue not to break with a friendship, even if the latter go against God." Finally the cleric stopped seeing the woman in question, only to die a year after Teresa had first met him. She sums up their history in this manner: "I never thought the great affection he bore me was wrong, although it could have been more pure."[62] Though Teresa refers to their relationship as a friendship, it has the hallmarks of an incipient love affair—one she was strong enough to resist. Teresa's "friendship" with the Lothario-esque priest presaged the modern nonsexual yet close relations that women form with men. In earlier times, society tended to regard such relationships with raised eyebrows, but even today the "just friends" designation can be fraught.

Ultimately, after years of systematic prayer, Teresa arrived at a state of mind in which she was able to reconcile her human friendships with her devotion to God. In her words: "I have never again been able to tie myself to any friendship or to find consolation in or bear particular love for any other persons than those I understand love Him and strive to serve Him."[63] Like her spiritual mentor, Saint Augustine, whose *Confessions* resonated with her like few other works, Teresa of Ávila came to regard friendship as meaningful and enduring only when it was cloaked in the love of God.

Like many of her male predecessors, Teresa was wary of close relationships that might be physically intimate. She even warned against the tactile gestures of affection that are so common among women: "Let no Sister embrace another or touch her on the face or hands." Instead, the sisters were enjoined to "include all in their love for one another, as Christ often commanded His disciples."[64]

In her later years, Teresa's close friends among her sister Carmelites included Ana de San Bartolomé, María de San José, and Ana de Jesús. The first Ana was a Castilian peasant who learned how to write in one of the numerous convents Teresa founded; she later became Teresa's personal assistant, secretary, confidante, and nurse. Under Teresa's guidance Ana also began to experience religious visions, which strengthened her faith. After the death of her "spiritual mother," Ana wrote the *Defence of the Teresian Legacy* in recognition of Teresa's unique contributions to the Catholic Church and, not incidentally, her own place as Teresa's spiritual heiress.[65]

Teresa's two other faithful companions came from more distinguished backgrounds. María de San José was a lady-in-waiting for an aristocrat before she professed vows, and Ana de Jesús came from a family of the lower nobility. Both helped Teresa spread the Discalced Carmelite Order in Spain and then abroad, María in Seville and Lisbon, Ana in Salamanca and France. Devoted to Teresa and committed to Church reform, these women and others offer models of friendships that served a cause greater than themselves.

### Beguines

Throughout Europe, other religious communities of women that did not require enclosure in a convent sprang up. The best known of these communities were the beguines, single or widowed women who lived together for spiritual purposes. The movement started in the Low Countries and eventually spread to Germany and France. Beguines shared a communal life in a house called a *béguinage*, in which each woman had her own apartment or room. All beguines were committed to chastity, poverty, charity, and prayer, though each group had its own guidelines. Unlike women in convents, most of whom came from noble families, beguines came from all levels of society and earned their modest living by taking care of the sick or working in the textile industry. Voluntary poverty and a commitment to saving souls were the hallmarks of these pious women, but the Church tended to treat them with suspicion because they were not bound by the regulations of an established order. After the Fourth Lateran Council of 1215 forbade the founding of new religious orders, the beguines encountered ongoing hostility from the Church, and one of them, Marguerite Porete, author of *The Mirror of Simple Souls*, was even burned at the stake for heresy in Paris in 1310.[66]

Because many of these women were literate and left behind letters, poetry, or treatises, we are able to get various glimpses of their lives. Some had joined the beguines because the two other options —marriage or the convent—were not possible for them: entry into marriage or a nunnery in the Middle Ages usually required a substantial dowry. Medieval society did not permit women to live together openly, but beguine communities offered female friends a reasonably respectable way to do so. Some, if they had a choice, preferred to live collectively with other women than to be confined within marriage or a convent. One German scholar has even credited these independent spirits with forming the first women's movement in European history.[67] Some beguine groups, especially those in France, were protected by the clergy, powerful patrons, or even the French crown, but beguines usually depended only on themselves to create communities dedicated to the love of God and their fellow creatures.

Like the beguines, other European women banded together for spiritual purposes without walling themselves off from the secular world. They were called by different names in different countries: *tertiaries* in Italy, *beatas* in Spain. The friendships made in these nonenclosed religious households served as ersatz families for women who had refused, or were denied, marriage. Some *béguinages* exist to this day in Belgium and the Netherlands.

## Sor Juana

By the late seventeenth century, convents were a staple not only of European society but also of the Spanish colonies in the New World. There, too, the social distinctions characteristic of society in general were mirrored in convents. Nuns with upper-class Spanish backgrounds were sent into cloisters with dowries, servants, and even slaves. While convents provided at least a basic education for novices from affluent families, this benefit was not offered to the servant nuns, much less the slaves. Moreover, convents tended to be segregated according to ethnicity and bloodline: there were convents for pure-blooded Spaniards born in Spain; for *criollas,* full-blooded Spaniards born in the colonies; for *mestizas,* of mixed Spanish and Indian blood; and for those with indigenous Indian or black backgrounds.

One Mexican woman who embodied the contradictions of class in New Spain, as well as its chance opportunities, was Sor Juana Inés de la Cruz (1648/51–1695). Juana began her unusual trajectory as the illegitimate daughter of a *criolla* mother and an absent, reputedly Basque, father. Raised on a farm outside Mexico City, the young girl learned to read within the aptly named *amiga* system, which encouraged local women to teach the rudiments of literacy to rural children. Juana proved to be an avid reader, and when she was sent to live with relatives in Mexico City she developed into a prodigy. Through the connections of a maternal aunt, she was placed at the court of the viceroy as a lady-in-waiting to the vicereine, Leonor Carreto.

Here she experienced the first of her two great friendships with women of superior rank who were sensitive to her intellectual gifts, talent as a writer, wit, and, not incidentally, exceptional beauty. Regarding her friendship with the vicereine, the Mexican poet Octavio Paz has written: "It was a relationship of superior to inferior, of protectress to protégée, but one in which there was also recognition of an exceptional young woman's worth." Their "friendship of the spirit" and shared love of the arts reminded Paz of other, more celebrated, male friendships.[68]

After four years in the company of the viceroy and vicereine, Juana chose to become a nun. Why she made this decision at the age of nineteen is still unclear, because she does not seem to have had a profound religious calling. But since she wanted the freedom to study and write, a convent was her best option. What we do know is that she first professed her vows at a strict Carmelite convent, but she later entered the more lenient Convent of San Jerónimo, a prestigious institution exclusively for *criollas*. Here she wrote the literary works that propelled her to fame, most notably the religious drama titled *El divino Narciso* ("The divine Narcissus").

Even after she had taken her vows, Sor Juana maintained close contact with one of the women who had befriended her at court: María Luisa de Laguna, countess of Paredes. As the new vicereine, María Luisa became Juana's most beloved friend and patron, and the subject of numerous passionate and witty poems. Because of Juana's growing stature as a writer and the permissive rules of her cloister, she received a great many prestigious visitors, including María Luisa and the ladies of her court. Meetings with outsiders should have been conducted through the wooden bars that separated convent inmates from the rest of the world, but exceptions seem to have been made for august visitors, who could be received in rooms called *locatorios*. Moreover, it is known that the viceroy and vicereine attended chapel at the convent and spent time afterward chatting with their protégée.

After eight years in Mexico María Luisa returned to the mother country, and she saw to it that *El divino Narciso* was presented at the Spanish royal court. She also oversaw the publication of the first of Sor Juana's collected works, *Castalian Inundation* (1690). Such exposure in Spain for the writings of a cloistered nun deep in the recesses of an unruly colony half a world away seems nearly miraculous today. María Luisa proved to be an exceptionally loyal friend to the remarkable woman she had left behind on the other side of the Atlantic.

Of the 216 poems gathered in Sor Juana's *Complete Works*, fifty-two, or almost a full quarter, are dedicated to "los marqueses de La Laguna," that is, María Luisa and her husband. These poems convey startling expressions of Juana's feelings for her aristocratic friend. Yes, one can argue that they follow the conventions of earlier poets writing in praise of their mistresses or patrons: Sor Juana does not hesitate to call the countess *Filis*, a name for the beloved commonly used by seventeenth-century Spanish and French poets. But in the next lines of the poem, which describe the particulars of Juana and María Luisa's circumstances, the poet considers their shared gender as unimportant to their enduring love as the geographical distance between them.

> *Your being a woman, your being gone:*
> *neither hinders my love for you,*
> *for you are well aware that souls*
> *ignore both distance and sex.*[69]

This may be one of the first poems ever in which the word *sex* is used as a marker of gender in reference to two women. In the original Spanish, the blunt word *sexo* startles, especially right underneath the ethereal *almas* (soul). Here and elsewhere, Juana's professed love for María Luisa jumps off the page as heartfelt. Still, she makes it clear that her feelings constitute "a pure flame" of adoration, one far removed from torments of the flesh.

In interpreting Juana's feeling for the countess, Octavio Paz speaks of a "neoplatonic" friendship between persons of the same gender, impassioned but chaste. Their relationship corresponded to the needs of the two women, one a virginal nun, the other a wife and mother. Each found in the other an outlet for emotions that were not fully satisfied by the religious or the conjugal life.

María Luisa seems to have been clever and lively, and endowed with a keen appreciation for literature. Finding in Sor Juana a sister spirit, she did everything possible to maintain their friendship during her eight years in Mexico, and she then continued it through correspondence when she returned to Spain. Alas, Sor Juana's letters to María Luisa have been lost, so once again we must depend on her poems for information about their continued relationship. In one poem Juana begs forgiveness for not having written; then she reaffirms her affection and speaks of the "blessing of your love," which indicates that her feelings for the countess were reciprocated.[70] To judge from the case of Sor Juana, and other nuns before her, emotions we usually associate with romantic love can also find a home in friendship.

In 2013, a play called *The Tenth Muse*, written by Tanya Saracho, premiered at the Oregon Shakespeare Festival.

The plot is set in motion when three young novices appear at the Convent of San Jerónimo in 1715, twenty years after the death of Sor Juana. They discover a cabinet containing a ribald play written by Sor Juana, and they decide to produce it. This play within a play leads to dramatic consequences for the three young women, who form tight bonds of sisterhood in the face of forces that would repress them. It seems fitting that their bonding with one another was inspired by the supposed works of Sor Juana Inéz de la Cruz, whose plays were indeed produced at the convent of San Jerónimo during her lifetime and whose history as a nun and author was so dramatically marked by her friendships with other women.

### Sisterhood of Like-Minded Spirits

The preceding examples of friendships formed by nuns, garnered from the twelfth through the seventeenth centuries, are obviously only a small sample from the bounty of relationships such women experienced. But they do tell us that nuns, like most human beings, were (and are) in need of sympathetic friends to share their burdens and joys. Our sample is weighted in favor of those nuns who gained renown as abbesses, mystics, and writers—women who had access to literacy and social connections. Whatever their circumstances, they reached out to the women cloistered with them and even to some beyond the convent walls.

Many nuns might not have chosen the convent without the pressure or explicit command of their families, since it was far cheaper to send a daughter to a convent than to hand her off, with a large dowry, to a suitable husband. For these nuns, halfhearted or even rebellious when they heard the convent door shut behind them, the friendship and instruction of other women probably eased their transition into cloistered life.

For women who chose the convent out of a true religious vocation, the sisterhood of like-minded spirits would have added to their happiness. As brides of Christ by choice, these women supported one another in their determination to lead exemplary lives—whatever the hardships—in the hope that life on earth would be followed by a beatific afterlife.

As the first large group of women in the Western world to experience adulthood outside the traditional family setting, these early nuns learned to form alliances as important to them as the ones they otherwise would have had with husbands, children, and other close kin. Having separated themselves physically from their families of birth, they found within the cloister a support system that gave them structure and meaning, as well as the "particular friends" that Church authorities sometimes found so alarming.

The general public went about its business unaware of this rich chapter in female friendship. For the most part, nuns' personal relationships remained enclosed within convent walls, like the nuns themselves, and had little influence on the ways European men and women conceptualized friendship.

# FOUR

# GOSSIPS AND SOUL MATES

*"We may generally conclude the Marriage of a Friend to be the Funeral of a Friendship."*

LETTER FROM KATHERINE PHILIPS TO CHARLES COTTERELL, 1662

O UTSIDE THE CONVENT, THE STORY of friendship continued to be written by men, with an exclusive focus on male subjects. The French philosopher Montaigne, inspired by his platonic love for Étienne de La Boétie, wrote his magisterial essay "Of Friendship," which quickly joined the works of Aristotle and Cicero to form a trio of foundation texts for anyone seriously interested in the subject. Other sixteenth-century humanists like Montaigne continued to conceptualize ideal friendship as an uplifting experience for virtuous men who were committed to one another through personal, religious, military, and civic bonds.

Yet during Montaigne's lifetime (1533–92), the reality of women's friendships was becoming more and more visible. An observer in 1580s England wrote that rich city wives sat before their doors finely dressed, that they went walking and riding and to play cards with other women, and that they spent time "visiting their friends and keeping company, conversing with their equals (whom they term gossips) and their neighbors, and in making merry with them at child-births, christenings, churchings and funerals; and all this with the permission of their husbands as such is the custom."[71] The word *gossip* was a common sixteenth-century term for a woman friend and had not yet acquired the derogatory connotation of idle talk or rumor that it has today. Gossips exchanged information that was personally and communally useful, and, in the case of bad behavior, their conversations constituted a means of enforcing social norms. If their concern about an incident became intense enough, town administrators might feel pressured into taking formal action against the transgressor. The word *gossip* was used in documents such as wills and court records, and the person of the gossip appeared regularly in plays performed on the Elizabethan stage.

### Female Friends in the World of Shakespeare

Shakespeare (1564–1616) bore ample witness to female friends in all social classes, ranging from the royal court to the alehouse. Mrs. Ford and Mrs. Page in *The Merry Wives of Windsor*; Queen Hermione and her faithful friend Paulina in *The Winter's Tale*; Hermia and Helena in *A Midsummer Night's Dream*; Mistress Quickly and Doll Tearsheet in *Henry IV, Part 2*; the French princess Katharine and her maidservant, Alice, in

*Henry V*; Rosalind and Celia in *As You Like It*; Cleopatra and her female attendants in *Antony and Cleopatra*; Beatrice and Hero in *Much Ado about Nothing*; Portia and her handmaiden, Nerissa, in *The Merchant of Venice*—these are only some of the girlfriends Shakespeare created as personifications of village gossips and high-born ladies who assisted one another to defeat male stupidity, misunderstandings, and outright violence. For the most part, the playwright not only confirmed the existence of female friends but also privileged their relationships, and he often resolved his plots through the agency of two women working in tandem to bring about a happy end.[72] Consider how Mrs. Ford and Mrs. Page in *The Merry Wives of Windsor* thwart Falstaff's efforts to seduce them. That bloated, drunken, charming reprobate knows so little of women's habits with their friends that he sends each the same written message, and, after they have compared their love letters, they devise a revenge that sends him packing in disgrace.

In *The Merchant of Venice*, Portia, disguised as a doctor of the law, and her handmaiden, Nerissa, as a lawyer's clerk, plead the suitor Antonio's case in court. The two women support each other so well in the ruse that they cleverly prevent Antonio from having to render the pound of flesh owed to the moneylender Shylock.

*The Winter's Tale* pits Queen Hermione and her friend Paulina against King Leontes, who has wrongfully accused his pregnant wife of adultery and sent her to prison. Paulina manages to bring Hermione's newborn daughter to the obdurate king to soften his heart, but he renounces her as a bastard and decrees that she should be exposed to the elements. Of course, since this is a comedy, in the end both the daughter and mother are saved. Ultimately they are reunited with a repentant king, all through the efforts of Hermione's loyal friend, Paulina.

In *As You Like It*, Celia and Rosalind are cousins and bosom buddies who have been bound by reciprocal affection since childhood:

> *We still have slept together,*
> *Rose at an instant, learn'd, play'd, eat together,*
> *And wheresoever we went, like Juno's swans,*
> *Still we went coupled and inseparable.*[73]

Eventually, however, their intimacy recedes in order to make space for heterosexual love, which often—in life as well as in literature—supersedes female friendship. The conflict between women's bonds with one another and marriage is one that we shall see played out, not only on the Shakespearean stage but also in real life, right into the twenty-first century.

The biographer Peter Ackroyd reminds us that Shakespeare's mother had six sisters and that she, the youngest, seems to have been the favorite in a community of women.[74] It is likely that Shakespeare imbibed an atmosphere of female solidarity from listening to his mother's conversations with her older sisters and female friends. Stratford, his childhood town, was a place where women, like men, depended on their neighbors for daily commerce and mutual support, and where married women in particular were valued as social agents, forming alliances among themselves that operated as a kind of unofficial police force.

One record (which implies that there may have been similar instances lost to history) describes how Eliza Neale was killed by her neighbor's husband as she intervened during his attempt to murder his wife. Were the two women friends? Given the epitaph on Eliza's gravestone, we can at least assume that they looked out for each other: "To save her neighbor she has spilt her blood / and like her Savior died for doing good."[75]

Especially before marriage, Elizabethan women spent most of their time with members of their own sex. About one in five never married, which represented a sizeable portion of the female population. Well into the twentieth century, as in Shakespeare's time, single women shared the same bedroom, even the same bed. Among working-class women, non-kin would often replace sibling bedmates as girls left home to take up apprenticeships, enter into household service, or find other jobs.

It was common for teenage girls in the sixteenth century to enter the service of gentry families in the countryside and afterward move into large London households. On average, they spent about four years working as a servant before marrying. Society did not tolerate single women living alone, as evidenced by a 1562 Act of Parliament, the Statute of Artificers, which required that all unmarried women be employed in service or sent to prison. Surely this extreme act was not intended for girls from middle- and upper-classes families.[76]

Working-class girls in service were bound by an awareness of their common condition and their need to sustain one another. It is not hard to imagine the intimate conversations that would have taken place as they whispered to each other after the candles had been blown out, comforting each other following a day of hard work, laughing at other people's foibles, revealing secret wishes, and sharing their creaturely warmth. Opportunities for intimacy would have been provided during the day as the two young women worked together on household duties such as cleaning, cooking, spinning, sewing, laundering, and making the beds, or as they walked home side by side from the market, the well, or church, or, during their time off, when they joined others of their age at local festivities.

Of course, quarrels between women would also emerge, as evidenced by court disputes involving individuals and families. Mature women in particular were called upon to bear witness against one another in trials concerning "inappropriate behavior" such as prostitution or witchcraft. Women witnesses often had the kind of detailed knowledge of one another's houses that only intimate access would afford; they moved without much announcement or formality among their neighbors' dwellings, and if something was wrong next door, a woman would know it and likely investigate. Sometimes baser motives, like jealousy, ignited the ire of one woman against her erstwhile friend and unleashed mutual recriminations over sexual wantonness that could turn into public fights. These incidents erupted spasmodically in friendship, as they did in marriage, disturbing the peace of a community and reminding people that even life's most cherished bonds are fragile.

Quarrels apart, women in the countryside, who washed their laundry on the riverbank or sold their vegetables in the marketplace, might have stopped along the roadside to exchange the latest gossip, complain of their husbands, or, tragically, lament the loss of a beloved child. Many women relied on their neighbors to come running in an emergency—for instance, when they went into labor, fell ill, or needed a substitute mother for their children. The lying-in period after childbirth—forty days for a girl, thirty days for a boy—afforded new mothers the opportunity to rest in bed, adjust to breastfeeding, and play with the baby, often in the presence of the midwife, female neighbors, and the baby's maternal grandmother, if there was one. Lying-in was "girl time" divorced from the demands of a husband, who was expected to abstain from intercourse with his wife and to pitch in to household chores. At the end of the period, the new mother would be purified, or "churched"—that is, brought to church with her baby, the latter in the arms of the midwife. Following the church ceremony, there would be a party in the mother's home, with cakes and ale furnished by her friends.[77]

Friendship with a neighbor often became every bit as important as biological kinship, since wives commonly moved away from their mothers and sisters and could rarely undertake the journey home, either by foot, horse, cart, or carriage. Those who knew how to write—probably no more than one in ten women—were at least able to stay in touch with their families through letters, but the great bulk of the female population was illiterate and had to count primarily on neighbors for companionship, comfort, and news.

Talk was the mainstay from which female friendship unfurled. Talking women—a subject for the ages! It is no accident that one's native language is called the mother tongue, *muttersprache, langue maternelle*. Mothers talk to their infants from the day of their birth, sing to them, and recite nursery rhymes, transferring an oral tradition that they themselves imbibed from their mothers and grandmothers. In the past, both boys and girls acquired speech primarily from their mothers or female caretakers during their first years of life. Thereafter, some of the boys—like Shakespeare—were sent to grammar schools, and an even smaller number of girls attended dame schools, where they learned the basics of reading, needlework, and, rarely, writing. In Protestant England, it was important for both boys and girls to be able to read the Bible, but writing was an option only for very few; in 1600, only about a third of men and 10 percent of women could write their names.

But talk—surely they all did that, with various regional and class accents that persist to this day. (Remember the Broadway hit and Oscar-winning movie *My Fair Lady*, based on Bernard Shaw's play *Pygmalion*.) Women in particular were deemed to be proverbial talkers, gossips, chatterboxes, prattlers, scolds, nags, rumormongers, and conversationalists—in short, verbally exuberant, in contrast to men in general and Englishmen in particular, who are still stereotyped as taciturn and emotionally reserved. While men excelled at public speaking and written expression, women were considered natural members of the social sex, given to friendly discourse and communal gossip.

What Englishwomen themselves thought and wrote about their condition as friends would become easier to discern during the seventeenth century, when they took up the pen in greater numbers and left behind a substantial body of writing, ranging from letters and diaries to poetry and plays. Two subjects that had been absent from most of the male-authored literature of the past would now come to the fore: woman as mother and woman as friend. Both of these subjects would find their way into the poetry of Katherine Philips, who championed friendship between women like no English author before her.

### Katherine Philips

Katherine Philips, née Fowler (1632–64), was the daughter of a wealthy London merchant and his second wife. At the age of eight she entered Mrs. Salmon's school in Hackney, where she met the first of her significant friends—Mary Aubrey, called *Rosiana* in Philips's poems. After her father died, his widow remarried and the family moved to Wales, and there Katherine, not yet seventeen, married James Philips, a widower of fifty-four. She gave birth to two children, a son who died at two weeks and a daughter who lived to bear sixteen children of her own. Though the marriage was apparently stable and happy, her significant relationships seem to have been with other young women.[78]

Even after she became a wife, nothing stopped her pen from flowing in praise of her cherished Rosiana:

*Soule of my soule! my Joy, my crown, my friend!*

. . . . . . . . . . . . . . . . . . . .

*I have no thought but what's to thee reveal'd,*
*Nor thou desire that is from me conceal'd.*
*Thy heart locks up my secrets richly set,*
*And my brest is thy private cabinet.*[79]

Katherine Philips emphasized the confidences that the two women freely shared. Best friend throughout the ages is often defined as the person to whom one can tell anything, even one's most private secrets.

When Rosiana married, Philips chose Anne Owen as her new best friend and gave her the sobriquet of *Lucasia* in her poems. Philips called herself *Orinda*, and that is what we shall call her from here on. In "On Rosiana's Apostasy and Lucasia's Friendship," Orinda bemoaned the loss of her schoolgirl friend through marriage, which caused her to take her soul back from Rosiana so as to send it on to Lucasia:

*Great Soul of Friendship whither art thou fled,*
*Where dost thou now chuse to repose thy head?*

. . . . . . . . . . . . . . . . . . .

*Then to the Great Lucasia have recourse,*
*There gather up new excellence and force.*

. . . . . . . . . . . . . . . . . .

*Lucasia and Orinda shall thee give*
*Eternity and make even Friendship live.*[80]

It was for Lucasia that Orinda wrote her most passion-ate poems. To her she exclaimed, "There's a religion in our Love." Thanks to Lucasia she never felt alone, because their minds were "so much one."[81] In another poem to Lucasia, Orinda expressed feelings that come as close to erotic love as was permitted for same-sex friends:

> *I did not live until this time*
> *Crown'd my felicity,*
> *When I could say without a crime,*
> *I am not Thine, but Thee.*[82]

She felt free to declare her love for another woman without fear of reprisal, since, as she put it, she and her friend were one—that is, one soul. Thus she was certain that no one could mistake her platonic vision for carnal longing.

Orinda had very distinct ideas about the nature of female friendship, based on the neoplatonic ideas made popular at the English court by Queen Henrietta Maria, the wife of Charles I. The concept of platonic love, derived from Plato's *Symposium* and embraced by sixteenth-century neoplatonic humanists, demoted eros to a mere stepping-stone in an ascent to divine love. It was understood that a woman could share the same soul with her best friend, but rarely, if ever, with her husband.

When Lucasia, too, decided to marry, Orinda was rightly fearful that she would lose her second soul mate. Although she found nothing to admire in Lucasia's choice of husband, she accompanied her to her new home in Ireland for what proved to be an emotionally unsatisfactory visit. That husbands sometimes stood in the way of female friendships was a common complaint among wives. Orinda expressed this conflict blatantly after Lucasia's marriage: "I find there are few Friendships in the World Marriage-Proof . . . we may generally conclude the Marriage of a Friend to be the Funeral of a Friendship."[83]

The enmity between marriage and friendship, as experienced by women, was by no means limited to the seventeenth century. We find evidence of it elsewhere in the English-speaking world—for example, in late eighteenth-century America, when Lucy Orr, writing to her friend Polly, exclaimed that marriage was "the bane of Female Friendship." She hoped it would not be the case with them "if we should ever Marry."[84] Jane Harrison, an eminent British scholar of Greek antiquity, added a personal lament in her 1925 *Reminiscences of a Student's Life*: "Marriage for a woman at least, hampers the two things that made life to me glorious . . . friendship and learning."[85] In today's pop culture, the novel *Bridget Jones's Diary* and the television show *Friends* have played upon the fear that marriage would disrupt cherished friendships. Recently in San Francisco, a thirtysomething single woman spoke in confidence about the changes taking place within her mostly married cohort and uttered this cri de coeur: "I'm hemorrhaging friends."

Though Orinda's visit to the Irish home of her newly married friend Lucasia was emotionally dispiriting, it proved productive in a different way. Orinda completed her translation of Pierre Corneille's play *La Mort de Pompée*, which made her immediately famous. She enjoyed only a brief period of literary glory, however, before dying from smallpox in 1664, at the age of thirty-two.

### Blue Stockings

Orinda's devotion to the ideal of friendship was more than a one-on-one commitment. In the early 1650s she founded the Society of Friendship, which included a coterie of like-minded women and a few select men.[86] Thereafter, other seventeenth-century English poets, such as Aphra Behn, Mary Mollineux, Jane Barker, Anne Killigrew, and Anne Finch, took up the baton and helped establish a model of loving friendship that British and American women would emulate in life and literature well into the late nineteenth century.

The subject of women's friendships spread beyond poetry, into the prose of the prefeminist philosopher Mary Astell (1666–1731), who explored what she saw as differences between men and women as friends. Astell believed that women were more capable of authentic relationships because they were less prone to be swayed by the dictates of self-interest that affected men in the public sphere. In her words: "As we [women] are less concern'd in the affairs of the World, so we have less temptation from Interest to be false to our Friend . . . our Sex are generally more hearty and sincere in the ordinary Friendships they make than Men, among whom they are usually clogg'd with so many Considerations of interest and Punctilio's of Honour."[87]

After Mary Astell came the Blue Stockings, a group of intellectual women who would gather in London for literary conversations under the leadership of Elizabeth Montagu (1718-1800). The term "Blue Stockings" is attributed to an incident that occurred when the scholar and botanist Benjamin Stillingfleet was invited to join Montagu's gatherings. Since he could not afford the black silk stockings worn by the ladies and came wearing his everyday blue stockings, the whole group came to be known as Blue Stockings. In time this label started to be used derisively to describe not only eighteenth-century British ladies but also any woman with intellectual aspirations.

Even more pernicious were the accusations of immorality against female friends that were first voiced in early-eighteenth-century scandal literature. Anonymous pamphlets likened women's friendships to the sexual practices associated with the ancient Greek poet Sappho, or the reputed intimacy of Turkish women in their harems.[88] Unlike Katherine Philips a century earlier, who had proudly proclaimed her platonic friendships, some women became extremely nervous about any situation that might incite negative talk. For example, one of the Blue Stockings was vexed that two women of her acquaintance were planning to live together. She wrote in a letter to her sister: "It will add to the jests the men made on that friendship, & I own I think those sort of reports hurt us all . . . I cannot think what Mrs L and Miss R can mean by making such a parade of their affection, they might know it wd give occasion to Lies."[89] However intellectual, genteel, and respectable she might have been, a woman known to frequent the Blue Stocking circle would find herself subject to suggestions of impropriety.

All the women discussed above had the financial means and leisure time to meet regularly in their homes with other similarly situated women. The doors were not closed to men, provided they were of the right sort—that is, men willing to recognize women's bonds with other women as second to none. These women were sophisticated and used to the amenities of urban life, foremost of which were the pleasures of sharing their lives with sympathetic friends.

### Friendships in the American Colonies

As we cross the Atlantic and look for evidence of women's friendships in the New World, we must remember that seventeenth-century America was both geographically and culturally isolated from the cosmopolitan centers of the mother country. To begin with, few female-authored texts of any sort can be found in American archives among the numerous records left behind by men, and there is certainly nothing resembling the paeans to friendship written by English women. The sole (remembered) American woman poet of the seventeenth century, Anne Bradstreet, used her considerable literary gifts to present herself as a loving wife and doting mother in the collection of her poems that was published in 1650, *The Tenth Muse*. By the time her book came out, she had lived in Massachusetts for twenty years and produced eight children. With her maternal and wifely duties, daily household responsibilities, and weekly church attendance, she certainly did not have the leisure time for friendship enjoyed by her British counterparts. In Puritan New England, it would have been ill advised for a woman to contest the absolute priority of marriage and family by lavishing attention, much less poetic paeans, on her friends.

Anne Bradstreet had arrived from England in 1630 with her husband, Simon, her parents, and her sisters. They had sailed on the *Arabella*, named for Anne's childhood friend Lady Arabella Johnson, who had also emigrated with her husband. Within months of their arrival, Lady Arabella and her husband had died. Losing her dearest friend was a severe blow to Anne, especially at a time when she was confronting a new land lacking in basic amenities such as decent shelter, let alone what she considered good manners. Fortunately she had her husband, her family, and the financial means to live comfortably in Ipswich and North Andover, Massachusetts, towns that were small enough that all the inhabitants knew one another.

Historian Laurel Thatcher Ulrich, in her groundbreaking book *Good Wives*, re-creates the conditions under which female friendships would have existed in colonial America. As in rural England, friends were almost always neighbors, forming "a community of women, gossiping, trading, assisting in childbirth, sharing tools and love, watching and warding in cases of abuse."[90] Proximity provided the starting point for most friendships, since women entered one another's houses without ceremony to borrow a range of items, such as scouring sand or an iron pot. They came running if they heard screams, and they intervened in quarrels between husbands and wives. They were there in great numbers to assist in childbirth, along with the midwife and the baby's maternal grandmother. Some offered their own breasts to nurse a newborn, since colostrum—the first milk produced by a new mother—was, mistakenly, considered unsuitable. They helped the new mother learn to breastfeed and they visited regularly until she was back on her feet, expecting that she would do the same for them when their time to give birth came. It was not unusual for a woman to give birth to as many as eight, ten, or even twelve children but lose half of them, either in childbirth or to childhood diseases. The friendly neighbors would be there to help the grieving mother clothe her infant for the last time before it was placed in a small wooden coffin and carried to the cemetery.

Proximity may have been the starting point for most friendships, but social rank was equally important. The mistresses of big houses, whose husbands were magistrates, ministers, or wealthy sea captains, socialized only with one another. Their social inferiors—the wives of small landowners and craftsmen—would also have found friends from their own social level. And the poorer women living in rented houses were likely to depend on one another for basic necessities as well as emotional support.

Despite differences in social strata, people from every level came in contact with one another. Farmers' wives sometimes hawked their produce door-to-door, which meant they would have chatted with their "betters" in the big homes not far from their own, more modest, dwellings. Poorer wives and their daughters would have kneaded bread in the kitchens of other women kind enough to lend them the warmth of a fire. Friendships surely formed as women worked together, gossiped, sang, shared secrets, and empathized in life's pleasures and trials.

Sometimes friends were asked to mediate love or marital affairs, performing for a woman something she could not do herself. Thus Sarah Woodward asked a friend to write a letter breaking off her engagement, although the wedding banns had already been announced. Presumably Sarah could not write, but her friend obviously could. Even with her friend's intervention the marriage was not called off, and Sarah Woodward rued her marriage to a man she did not love.[91]

Compared to the London literary ladies of the seventeenth century, who supported one another both individually and collectively, female friendship in America was still a one-on-one phenomenon. It would take another century or more for American women to catch up with their English counterparts.

# FIVE

# PRECIOUS LADIES

*"When people feel tender friendship, it is so sincere and ardent and intense that all the sorrows and all the joys of their loved ones feel like their own."*

MLLE DE SCUDÉRY, *CLÉLIE*, 1654-61

*"It is more shameful to distrust our friends than to be deceived by them."*

FRANÇOIS DE LA ROCHEFOUCAULD, MAXIM NO. 84, 1665-78

I N THE SEVENTEENTH CENTURY, DURING the reigns of the French kings Louis XIII and Louis XIV and the English kings Charles I and Charles II, the cultural ties between the two nations were very close. In 1625 the sister of Louis XIII, Henrietta Maria, came to England to marry Charles I; she brought with her a large French entourage and the cult of platonic love already popular in certain elite French circles. English queens, such as Henrietta Maria and her more famous predecessor, Elizabeth I, and French queens, such as Marie de Medici and Anne of Austria, maintained an entourage of ladies-in-waiting bound together by loyalty to their sovereign mistresses. Whatever rivalries they experienced (and these were legion!), ladies-in-waiting set a pattern of friendship that was widely envied and emulated by their peers.

If the court set the gold standard for socially and politically useful friendships, the city proved even more significant for women determined to establish close relationships beyond their families. Literary circles in London and salons in Paris provided a space for friendships to flower between women and, to a lesser extent, between women and men. Initially a Parisian phenomenon, the salon was eventually imitated in cities throughout France and then throughout Europe. Indeed, one could argue that French literary salons were the ancestors of all the women's clubs that have come into being since then: Blue Stocking conversations in eighteenth-century England; romantic salons in nineteenth-century Germany; and American book clubs, garden clubs, suffrage clubs, Junior League, Hadassah, and the like.

### The Marquise de Rambouillet's Weekly Salon

Just as Henrietta Maria brought refinement to the English court in the decades before the Civil War of 1644 and the execution of her husband in 1649, so too, during the same period, was French society enhanced by the brilliant salons hosted in the Parisian home of the Marquise de Rambouillet. Her weekly salon was the first in France where women participated equally with men and where they imbued social life with a distinctly feminine flavor. Regular attendees included the future writers Mlle de Scudéry, Mme de Sévigné, and Mme de La Fayette, as well as already established male writers such as Chapelain, Corneille, and Ménage, and members of polite society without literary ambitions of their own. Mme de Rambouillet encouraged them all to elevate their speech and behavior to a level considered appropriate for refined upper-class society. Because many of the women avoided words they considered vulgar and replaced them with euphemistic expressions, they all came to be labeled *précieuses* (precious ladies), with little distinction between those who embraced the affectations of preciosity and those who did not. Unfortunately, Molière's play *Les Précieuses ridicules* (1661) heaped ridicule on all of them, so that for most French people today *precious ladies* is automatically followed by *ridiculous*.

Some of the verbal refinements newly in vogue did lend themselves to derision. In *Le grand dictionnaire des précieuses* by Somaize, a dictionary of expressions reputedly used by *précieuses*, the moon was called "the torch of silence," tears "the daughters of pain and joy," the tongue an "interpreter of the soul," and death "the all-powerful." Instead of saying "Please sit down," one might have ventured: "Be so good as to accept the desire that this chair has to embrace you."[92] Whether these or other euphemisms were actually heard within Mme de Rambouillet's Paris residence is still a matter of scholarly debate.

More to our point, these women consciously set about changing notions of how one should act as a friend or lover.

Indeed, the very word for friendship—*amitié*—broadened its scope and evolved to include the tender emotions that lovers, as well as friends, were expected to feel for one another. Friendship, as the *précieuses* conceived it, was a noncarnal union of similar souls that could be experienced across genders. On rare occasions, friendship could even cut across rank, regardless of the colossal difference in wealth and status between a super-rich member of the nobility, like the marquise herself, and a talented writer with few financial assets, like Mlle de Scudéry. In time, even some members of the bourgeoisie were accepted within *précieuses* circles. Still, however far it spread into bourgeois or provincial settings, preciosity was predominantly an upper-class Parisian phenomenon, its rules for speech, dress, and polite behavior to be acquired over time in a sophisticated milieu.

### Mlle de Scudéry

The various personalities who assembled in the Blue Room at the Hôtel de Rambouillet were deftly portrayed under pseudonyms in *Artamène, ou Le Grand Cyrus*, a ten-volume roman à clef authored by Mlle de Scudéry. She presented Mme de Rambouillet, under the name of Cléomire, as a woman admired for her beauty, wit, nobility of spirit, generosity, taste, judgment, and high standards, all of which made her both feared and respected. Cléomire paid social calls less frequently than other Parisian women of her milieu, yet she was always surrounded with company since "there was no one in all the court with a semblance of wit or virtue who didn't go to her residence."[93]

The marquise and her female friends set the tone for both genders, the women no less talkative than the men, no less able to engage in the literary discussions that pitted some members against others and sometimes tested the limits of civility. After a heated quarrel over a certain Italian comedy (the Italian Ariosto's *I Suppositi*), Mlle de Scudéry tried to smooth things over with the marquise's daughter, Mlle de Rambouillet, who did not share her own positive opinion of the play.[94] It was essential for Mlle de Scudéry to maintain good relations with both mother and daughter if she wanted to remain inside the coterie admitted to the Blue Room.

Elsewhere in *Artamène* Mlle de Scudéry praised this same mademoiselle, under the name of Philonide, for her conversation, writing, knowledge, dancing, general charm, and social ease at court. But, a bit slyly, she also raised questions about the young woman's facile relationships with too many friends. It's a passage worth considering:

> Moreover, she has a multitude of female friends and male friends [*amies et amis*], not to mention her suitors, so prodigious that one wonders how she can respond simultaneously to the friendship of so many people . . . I am persuaded, no matter what she says, that it is not possible for her to love so many people . . . I'm sure there are a large number for whom she feels only respect, civility, and a certain gratitude. However, people are happy with her and love her as if she truly loved them.[95]

Can one have too many friends? Mlle de Scudéry, like Aristotle, seemed to think so. Try as she might to flatter this daughter of the imposing marquise, Mlle de Scudéry hinted at a certain inauthenticity in the practice of showing the same friendly face to everyone.

Satirized as affected, pedantic, and prudish, the precious ladies of seventeenth-century France were nonetheless instrumental in promoting what can now be seen as a protofeminist agenda. The social life they created for themselves, with relative independence from their husbands, allowed them to form friendships with both women and men and to pursue cultural activities that had previously been dominated by men. It was no small matter that they invented the French salon, which started out female-centered and would remain so for the next 350 years. And because there was a place for literary women like Mlle de Scudéry to try their wings, it was even possible for some of them to thrive outside of marriage or the convent, with friendship providing a network of support.

Mlle de Scudéry never married. Instead, after the success of *Artamène,* she established her own literary salon at her house in the fashionable Marais district of Paris. There, her friends included not only aristocratic women and male writers who had frequented the Hôtel de Rambouillet but also bourgeois women from her neighborhood (Mme Bocquet and Mme Arragonnais, among others). Many of these friends appeared, once more under pseudonyms, in her second meganovel, *Clélie,* which became a bestseller in France and throughout Europe.

Nobody reads Mlle de Scudéry's novels today—they are too long, too "precious." As the eponymous heroine of *Clélie* observes, "I have never heard anyone speak of a tender love, and I always imagined that this affectionate and significant term was consecrated to perfect friendship, and that it was only in speaking of it that one could employ the word tender." According to Clélie, it is tenderness that makes us able to see from another person's point of view—an ability we call *empathy* today. It is tenderness that makes us prefer spending time with an unhappy friend to a more entertaining pastime. It is tenderness that makes us excuse our friends' faults and exaggerate their virtues.[96]

If pressed for an answer, Mlle de Scudéry would have placed friendship above erotic love. She and others of her ilk had good reason to be wary of the fickle gallantry that passed for love among so many of their contemporaries. If they were single like Mlle de Scudéry, widowed like Mme de Sévigné, or physically and emotionally separated from their husbands like Mme de La Fayette, they might find in friendship what they had not found in marriage—a soul mate not subject to carnal desire.[97]

### *Mme de Sévigné and Mme de La Fayette*

Among the numerous female friendships that thrived in upper-class society, that between Mme de Sévigné and Mme de La Fayette was exemplary. They knew each other for more than forty years, were in and out of each other's homes almost daily when they were in Paris, kept in touch through the post when they were apart, and supported each other through good times and bad. They were helpful to their respective family members as well, and, like so many other courtiers during the reign of Louis XIV, never missed an opportunity to promote the well-being of their loved ones through their personal connections with high-ranking officials and the king himself. As in other historical eras, what Aristotle called a "friendship of utility" was for Mme de Sévigné and Mme de La Fayette intertwined with authentic, mutual caring.

Their friendship was also exceptionally well documented, since each woman was a consummate writer: Mme de Sévigné became the best-known letter writer of her day and Mme de La Fayette the most accomplished novelist, even though her books were published anonymously during her lifetime. Though they each married, bore children, and were in constant contact with many other members of their social milieu, they came to care for each other as best friends. Indeed, toward the end of her life Mme de La Fayette wrote to Mme de Sévigné: "Believe me, my dearest, you are the person whom I have truly loved the most in all the world."[98]

Mme de Sévigné (1626–96) was already a married woman with two small children when she first met Mlle Marie-Madeleine Pioche de La Vergne, the future Mme de La Fayette (1634–93). Eight years older, comfortably settled in the Marais where she had grown up, Mme de Sévigné was already a regular at the Hôtel de Rambouillet and on close terms with some of the better-known writers of her day, most notably the erudite poet Gilles Ménage. The two women became distantly related through marriage in 1650, but even if they had not shared some of the same relatives, they were destined to become friends, since they belonged to the same social and literary circles and sympathized with each other from the start.

### Madame de Sévigné, Mme de La Fayette, and Ménage

Like Mme de Sévigné, Marie-Madeleine struck up an important friendship with the poet Ménage. Having female friends did not preclude having male friends as well, especially in French upper-class circles, which never segregated the sexes to the extent found in many other countries. In 1650, when Ménage was thirty-eight and Marie-Madeleine almost seventeen, he was happy to offer her the same gallantries he had lavished upon Madame de Sévigné. As an abbot, though not a priest, Ménage was free to spend time courting women in the platonic mode that had become fashionable in France and England. In 1651 Mme de Sévigné was especially sensitive to his favors, since her husband, a notorious ladies' main, had gotten himself killed in a duel over his mistress's honor.

Ménage once tried to clarify his relations with them in a note to the writer Huet: "I think you have heard me say in the past that I loved Mme de La Fayette in verse and Mme de Sévigné in prose."[99] It seems that his relations with Mme de Sévigné were always solid and reliable, like prose, whereas those with Mme de La Fayette were more romantic and sensitive, like poetry. To take only one example of their mutual recriminations, Mme de La Fayette complained that Ménage sometimes did not tell her things that were common knowledge among their friends: "Being your friend to the extent that I am, it is ridiculous that I am always the last to know the things that concern you, and that I am ashamed to let others know I am ignorant of them."[100]

Fortunately for posterity, Ménage saved a great number of Mme de La Fayette's letters, but, unfortunately, most of the ones she wrote to Mme de Sévigné have been lost. Still, the bulk of Mme de Sévigné's own correspondence, directed from around 1670 onward to her married daughter, Mme de Grignan, constitutes a literary and historical treasure. These letters were so lively, so detailed, so entertaining that they circulated in numerous copies. Mme de Sévigné's cousin, admirer, and friend Roger de Bussy-Rabutin, summed up the feelings of one who received her letters with delight: "Yesterday I received your letter, Madame; it is five pages, and I assure you that I found it too short . . . there are, it seems to me, charms in your letters that one does not see elsewhere, and it is not the friendship I bear you which embellishes them for me, because many discerning people who do not know you have [also] admired them."[101]

After Marie-Madeleine Pioche de la Vergne became Mme de La Fayette in 1655, she and Mme de Sévigné were unable to see each other as easily as they had before. The Count de La Fayette, a widower almost twice the age of his bride, had properties in far-distant Auvergne, where the couple would live for much of the year. But Mme de La Fayette managed to return to Paris for long visits, and eventually she was able

to reside there permanently in her home on the rue de Vaugirard, while her husband stayed in Auvergne to administer
his estate. It was not an unusual arrangement for members
of the nobility during the seventeenth and eighteenth centuries, when marriages were affairs of family name and finance
rather than the heart. For aristocratic women, such an arrangement had the advantage of allowing them greater freedom to pursue an active social life with friends of both sexes.

On her return to Paris, Mme de La Fayette renewed her
face-to-face encounters with Mme de Sévigné and the poet
Ménage. She counted on both of them for help when she was
obliged to look for Parisian lodgings other than her own house,
which had been rented out. She instructed Ménage explicitly:
"I would very much like to lodge near Mme de Sévigné, that
is, near the Place Royale [today the Place des Vosges]."[102] In
the end she moved back to her own house, on the rue de Vaugirard, where she gave birth to a son in March 1658. After
the birth, as was the custom among women of her class, she
received Mme de Sévigné, Ménage, and select friends in the
space between the bed and the wall, called *la ruelle*. Because
bedrooms of affluent aristocratic women were often very
large, a sizeable number of friends could fit in *la ruelle*, and,
in time, the word became synonymous with a literary salon.

Once she was back up and about, Mme de La Fayette
resumed her connections with the *précieuse* group that gathered regularly at Mlle de Scudéry's "Saturdays" in the
Marais. She also began to frequent the royal court episodi-
cally, through her friendship with Henriette d'Angleterre

(Henrietta of England), the wife of Louis XIV's brother. This friendship had begun when Henriette was a girl, having joined her French-born mother, the English queen Henrietta Maria, in exile after Charles I was beheaded. Though Henriette was ten years younger than Mme de La Fayette, the two women nonetheless felt a kind of spiritual kinship that was to last until Henriette's unexpected demise at the age of twenty-six. Deeply moved by this tragic death, Mme de La Fayette wrote Henriette's biography (*Histoire de madame Henriette d'Angleterre*), which the young princess had herself once suggested. It was intended as a private memoir and remained so during Mme de La Fayette's lifetime.

Mme de La Fayette's first published work was a portrait of Mme de Sévigné.[103] Writing in the guise of *un inconnu* (an unknown man), Mme de La Fayette sang the praises of an unnamed woman, known by all to be Mme de Sévigné: "Your soul is large, noble, capable of dispensing treasures . . . You are sensitive to glory and ambition, and you are no less sensitive to pleasure . . . Joy is the true state of your soul, and chagrin is more out of tune with you than with any other person in the world."

The author, still posing as a man, goes on to say: "You are naturally tender and passionate, but to the shame of our sex, this tenderness has been useless to you, and you have contained it within your own sex by giving it to Madame de La Fayette."[104] This is an amazing text! Mme de La Fayette let it be known to all the world that she, not a man, ruled the heart of her dearest friend. She publicly affirmed their friendship at a time when *amitié* was at the height of fashion and could pertain to same-sex as well as heterosexual attachments.

Like the emotions expressed in the English poetry of Katherine Philips, the feelings Mme de Sévigné had for Mme de La Fayette were primary, with the exception of her obsessive love for her daughter, Françoise-Marguerite, born in 1646 and known to posterity as Mme de Grignan. As the major recipient of Mme de Sévigné's lengthy epistles, published in three fat volumes in the prestigious twentieth-century Pléiade edition, Mme de Grignan has a unique place in French literary and social history.

To have her daughter near her and to be surrounded by her closest friends was Mme de Sévigné's idea of paradise, as during the summer of 1667, when they were all in the countryside. She wrote to a friend:

I have M. d'Andilly at my left hand, that is to say on the side of my heart. I have Mme de Lafayette at my right; Mme du Plessis in front of me, who amuses herself by sketching little images, and Mme de Motteville a little further away, who is profoundly dreaming; our uncle de Saissac, whom I fear because I scarcely know him; Mme de Caderousse, his sister, who is a new fruit whom you do not know, and Mlle de Sévigné all over the place, coming and going like a little hornet.[105]

Elite women like these had the means to make of friendship a "movable feast" that could frequently be enjoyed in their country properties, at a great distance from their city homes. Unlike the lower classes, they were not bound by proximity to restrict friendship to their neighbors.

But soon the time came when it was necessary for Mme de Sévigné to give her daughter away in marriage. Quick to support this endeavor, Mme de La Fayette lent Mme de Sévigné the considerable sum of five hundred *livres* for her daughter's dowry—surely a mark of true friendship, since it would not be repaid for several years. In 1669 Mme de Sévigné's daughter married Monsieur de Grignan—a forty-year-old, twice-widowed count—who was subsequently named governor of Provence. This new position meant that his wife was obliged to accompany him to the south of France. Mme de Sévigné was inconsolable. It was the tragedy of her life, mitigated only by the companionship of her devoted friend on the rue de Vaugirard.

### Mme de La Fayette and François de La Rochefoucauld

By this time Mme de Lafayette had also begun her close friendship with François de La Rochefoucauld, the distinguished author of memoirs and maxims. This heterosexual friendship was emblematic of the new relationships available to some aristocratic women by virtue of their efforts to become the intellectual equals of men. No one could deny them a seat at the center of culture when they, too, could read and write and discuss literature, art, and music. And since women organized the salons, it was in the interest of male writers to be on good terms with their hostesses and their female entourage. Published authors like La Rochefoucauld and Ménage had no reservations treating Mme de La Fayette and Mme de Sévigné as worthy discussants and even potential writers, though it was understood that women of their rank would publish only anonymously.

Just as Mme de Sévigné had "shared" Ménage with Mme de La Fayette, so too did Mme de La Fayette "share" La Rochefoucauld with Mme de Sévigné. Despite his famously caustic vision of humanity, Mme de Sévigné came to count on him for another sympathetic ear in her constant efforts to mitigate the negative effects of her daughter's absence. In order to keep Mme de Grignan always present in their lives, she gave Mme de La Fayette a fine portrait of her daughter.

When Mme de Sévigné left Paris to visit her daughter in Provence, saying good-bye to Mme de La Fayette proved extremely painful: "Mme de Lafayette's delicate nature cannot easily suffer the departure of such a friend as I am."[106] By *delicate*, she was alluding not only to Mme de La Fayette's refined sensibility but also to the variety of maladies she habitually endured, namely "vapors" and fevers, which worsened as she got older.

For her part, Mme de La Fayette never lost an opportunity to say something nice about Mme de Grignan in her letters. On July 14, 1673, she wrote to Mme de Sévigné in Provence: "Please kiss Mme de Grignan for me and all her perfections."[107]A few months later, when she heard that Mme de Sévigné was delaying her return to Paris, she wrote: "Provided that you bring back Mme de Grignan, I won't complain."[108]

Being a friend to the mother meant being a friend to the daughter—and even to the son, Charles, whom Mme de Sévigné tended to neglect, just as Mme de La Fayette neglected her own sons. Their woman-centered friendship included Mme de Grignan as a "third self," but not the male relatives—neither sons nor husbands. After her short, unsatisfactory marriage, Mme de Sévigné never wanted to marry again, and Mme de La Fayette's husband was so invisible that when he died in 1683, it was as if he slipped away unnoticed.

La Rochefoucauld provided a masculine presence. The "couple" formed by La Rochefoucauld and Mme de La Fayette, based on their literary interests and their similarly poor health, easily expanded to form a threesome whenever Mme de Sévigné was with them in Paris. When she was away in Provence, Mme de La Fayette kept her apprised of her and Rochefoucauld's Parisian activities, always letting her know how much she was missed: "I am hoping for your return with the kind of impatience that is worthy of our friendship."[109] On the day of her return to Paris, Mme de Sévigné was greeted by a bevy of friends, including Mme de La Fayette and La Rochefoucauld, and when she subsequently retired to her room to recover from her journey, La Rochefoucauld remained on guard to make sure she stayed there for two days.

Together again in Paris, the two women saw each other constantly, both at their homes and at dinner parties hosted by their friends. Sometimes they went together to musical events, such as the opera, which would bring tears to their eyes.[110] They also spent time in the nearby countryside. Since afternoons and evenings were almost always reserved for her friends, Mme de La Fayette must have found time to write only in the morning.

In 1678, *The Princess of Cleves* was published anonymously. Though Mme de La Fayette never admitted publicly to being its author, it was rightly assumed that she had written it, perhaps in collaboration with La Rochefoucauld. The novel became an instant bestseller in France, and it was followed the next year by a popular translation into English. As the author's closest friend, Mme de Sévigné was thrilled with the novel and made sure that everyone she knew, including several priests, read it.[111] It would have been impossible for her not to have known the identity of the anonymous author.

Soon, however, Mme de Sévigné had to report to her daughter the sad news that La Rochefoucauld was dying. For days, she spent "almost all her time at the home of Mme de La Fayette, who would poorly understand the delights of friendship and the tenderness of the heart, if she were not as afflicted as she is."[112] Two days later, La Rochefoucauld was dead. Mme de Sévigné shared the affliction of her "poor dear friend," and asked rhetorically: "Where will Mme de Lafayette find another such friend?" She observed astutely that the long-term sicknesses of both parties had "rendered them necessary to each other," and she sincerely believed that "nothing could be compared to the confidence and charm of their friendship."[113]

Mme de La Fayette was to live another thirteen years. Fortunately, she could always count on Mme de Sévigné. In old age, both drew their sons closer to them, and, after an estrangement of twenty years, Ménage returned, reminding them of the gallant attentions they had enjoyed in their youth. Mme de La Fayette was the first to break the silence with him: "I would like to have news of you, monsieur. Our old friendship has not left me without concern for your health."[114] After Mme de La Fayette and Ménage had reconciled, she told him the special meaning that friendship had for her in her declining years: "I want to tell you how much I am truly touched by your friendship . . . Time and old age have taken away all my friends."[115] Friendship matters, especially in old age, when death reduces the number of one's friends.

Mme de Sévigné continued to convey a flow of news to her daughter, as if she were the official chronicler of Mme de La Fayette's life. In one frequently quoted epistle, she wrote these striking words: "Mme de Lafayette is rich in friends from all sides and all conditions. She has a hundred arms. . . ."[116] Mme de La Fayette did indeed have many friends, but when Ménage died in July 1692, she never got over her grief and was mostly bedridden until her own death on May 25, 1693.

The friend who meant the most to her—Mme de Sévigné—was in Provence at the time and could mourn her only from a distance. The surviving woman poured out her sorrow in a long letter to another member of their social circle: "You know the merit of Mme de Lafayette . . . I found myself only too happy to be loved by her for a very considerable period of time. We never had the slightest cloud in our friendship . . ."[117]

Mme de Sévigné remained in Provence for three more years and died there, with her daughter at her bedside.

As a result of the relatively high level of literacy among women from the French and English upper classes, records of their friendships have survived in troves of letters, memoirs, poetry, and novels. It is true that most of these pertain primarily to a small percentage of the population—to upper-crust women who had little to do other than visit with and entertain one another. Social position and money were necessary conditions for entry into this world, and shared leisure seems to have been the sine qua non for maintaining friendships over time. Nevertheless, these were true friendships in every sense of the word. Individuals chose other individuals on the basis of proximity, personal attraction, a sense of likeness, and similar interests. Undoubtedly there were many friendships initiated by self-interest and reciprocal services—motives that have been recognized by male thinkers from Aristotle to La Rochefoucauld—but this did not prevent friends from caring for one another in a more selfless manner as well. Witness Mme de Sévigné and Mme de La Fayette, who loved and aided each other for more than forty years.

The seventeenth century represents a turning point in the story of women's friendship. In England and France, within upper-class circles, women assumed an ascendant role in the making and breaking of friendships. Just as they had played a major role in the French invention of courtly love five hundred years earlier, so, too, would *salonnières* become directors of social life for elite women and men.[118] Led by notable hostesses, these denizens of high culture sharpened their social skills and developed the refined speech and behavior that would be de rigueur in polite society for years to come. In time, what transpired in England and France at the uppermost levels of society reached across Western Europe and, somewhat later and more democratically, even to select circles in colonial America.

# SIX

# PATRIOTIC FRIENDSHIPS

*"Can patriotism dwell in a heart where friendship has no place?"*
—Catherine Macaulay to Mercy Otis Warren, 15 July, 1785

*"Old friendships can never be forgotten by me."*
—Abigail Adams to Mercy Otis Warren, 4 March, 1797

THE BONDS OF FRIENDSHIP FORMED during political up-
heaval and war are among the strongest experienced
by humankind. Men, and more recently women too,
often remember their military comrades with a kind of loyalty
rare in a lifetime, even if they do not continue to see one another on
a regular basis afterward. Similarly, political causes can become
catalysts of friendship, with participants pairing off to become permanent twosomes. Even after the cause has faded, friends treasure
their shared involvement in a movement that gave them a collective sense of identity and brought meaning to their lives.

During the eighteenth century, while most French and British subjects supported the rule of monarchy, a republican sentiment took root in colonial America. By the 1770s, American advocates of rule by the people or their elected representatives were making their voices heard. Originating in "the soft whispers of private friendship," in the words of Mercy Otis Warren, the cry for freedom would eventually be heard around the world.

Mercy Otis Warren, an erudite writer living in Plymouth, Massachusetts, became the central figure in a circle of women friends drawn together by republican sympathies. Like their male counterparts, they looked back to ancient Roman heroes and heroines and constructed for themselves a model of dedication to the ideals of patriotism and liberty. They thought of themselves as participants in the making of an exemplary republic, for which they were willing to make sacrifices that were appropriate to their sex. They boycotted imported tea as a protest against British taxes, signed oaths of loyalty to patriotic causes, and raised money for military defense.[119]

In the area surrounding Boston, patriotic women communicated mainly through letters, with occasional visits to one another's homes. Many of these women came from well-off families and had received relatively good educations at home, even if they hadn't gone to Harvard like their brothers. A special feature of their friendships was that their husbands were also involved. Some letters were even written collaboratively by both spouses, and visits were often family affairs, with husbands and children in tow.

Mercy Otis Warren was descended from a family of politicians, among them her father and brothers. Her female correspondents had similarly distinguished pedigrees: For example, Hannah Winthrop was the wife of the Harvard mathematician John Winthrop, whose forebears were among the Massachusetts Bay Colony founders. Abigail Adams was the granddaughter of the gentleman farmer and politician John Quincy and wife of the future president John Adams. Furthermore, Warren's epistolary reach stretched across the Atlantic to the famous British historian Catharine Macaulay, a known sympathizer with the American cause. Because Mercy Warren, Hannah Winthrop, and Abigail Adams were married to very prominent men, and because Warren and Macaulay were published authors, their correspondences have been preserved to a remarkable degree. Remember that letters, especially those connected to public and semipublic figures during the revolutionary years, were meant to be shared, read aloud, and circulated in numerous copies, not unlike those of Mme de Sévigné in France a hundred years earlier.

### Mercy Otis Warren and Abigail Adams

Mercy's friendship with Abigail Adams began when the Adams couple paid a summer visit to the Warrens in Plymouth, and the friendship was subsequently sustained primarily through correspondence. As Abigail Adams's biographer Edith Gelles has written:

For the ensuing forty-one years, these two exceptional women regarded each other as friends, bonded by social and religious background, loyalty and empathy as women, and mutual regard for each other's intelligence. They did not always agree on issues; they frequently debated, and for long periods of time they would become estranged. The bond held, however, mostly because both of them willed it to survive.[120]

With a sixteen-year age difference between them and Mercy already something of a local literary star, their friendship began on unequal terms. If at first Abigail looked up to Mercy as a mentor, in time, as Abigail became more sure of herself, she learned to hold her own against her formidable friend.

Each woman saw herself as a patriot comparable to her prominent public servant husband, albeit in the domestic realm, where their responsibility lay in rearing their children (five for Mercy, four for Abigail) to become virtuous citizens. Each was concerned with her role as wife and mother, but unlike Anne Bradstreet in the seventeenth century, these eighteenth-century women engaged in discussions about public affairs once considered exclusively the province of men. Thus, soon after their initial encounter Abigail wrote to Mercy about the now-famous Boston Tea Party, at which, in protest against taxes levied by the British Parliament, patriotic Americans dumped an entire shipment of tea into Boston Harbor on the night of December 16, 1773. Abigail, like other resentful American colonists, understood correctly that this act of rebellion would lead to much greater devastation in the name of liberty: "The Tea that bainfull weed is arrived. Great and I hope Effectual opposition has been made to the landing of it . . . The flame is kindled and like Lightning it catches from Soul to Soul."[121]

Throughout the American Revolutionary War, Abigail confided to Mercy her impassioned hopes and fears for the revolutionaries, and Mercy responded in kind, though her prose style was characteristically stiffer and more didactic. During the years when their husbands were actively engaged in the revolutionary struggle, shared patriotism and religious beliefs strengthened their friendship. When John Adams became a founding father in Philadelphia, leaving his wife behind to manage the farm in Braintree, Abigail found a sympathetic ear in Mercy, whose own husband was a member of the Massachusetts House of Representatives, which was situated a long distance from their home in Plymouth. The much greater distance between Philadelphia and Braintree—at least a two-week journey—meant that Abigail would see little of John for the next two years.

The two women were somewhat consoled by their belief that they had revived the virtue of ancient Roman matrons through their own patriotic ideals and sacrifices. While American men looked to the Greco-Roman soldier and civic hero for their ideal, women found their role model in the Roman matron—a married woman who was reputedly chaste, dignified, self-sacrificing, and learned.[122] To strengthen this connection, Mercy proposed that Abigail assume the pen name Portia, the wife of Brutus, while she called herself Marcia, probably in reference to the wife of Cato the Younger. (We have seen the literary use of sobriquets before, in the poetry of Katherine Philips and the novels of Mlle de Scudéry.) But later on, when Abigail complained that she was suffering because of her husband's absence, Mercy dismissed her friend's loneliness with a self-pitying response: "You have sisters At Hand and Many Agreeable Friends around You which I have not. I have not seen a Friend of an afternoon Nor spent one abroad Except once or twice I Rode out since I Came from Braintree."[123] Mercy was referring to her recent visit to Abigail in Braintree, where she had observed Abigail surrounded by sisters and friends. Burdened by many family difficulties, including the loss of fortune and several sons' war injuries, Mercy was becoming unable to function as a sympathetic friend.

During the Continental Congress of 1776, an emboldened Abigail wrote to John her now famous letter asking that he "Remember the Ladies." Reminding him that "all Men would be tyrants if they could," she suggested that women, like men, should benefit from the Revolution and be protected by laws to restrict patriarchal rule. When John treated the matter as a joke, Abigail vented to Mercy: "I ventured to speak a word in behalf of our Sex, who are rather hardly dealt with by the laws of England which give such unlimited power to the Husband to use his wife Ill." She wished that some laws would be established "in our favour upon just and Liberal principals," and even proposed that women should make a collective protest. No letter has survived to indicate whether Mercy shared Abigail's feminist sentiments.[124]

Famously, Abigail and John addressed their letters to each other with the words "Dearest Friend" during a marriage that lasted more than fifty years. Theirs was indeed a companionate union for whom the word *friend* had real meaning. They saw in each other a complementary self who would always care for the other's well-being, even when the only means of sharing their feelings was through letters.

And so, at the end of 1777 when John was appointed to the delegation that would negotiate a treaty of alliance with France, Abigail could not endure the idea of a transatlantic separation. She wrote to Mercy hoping for sympathetic understanding, but instead her friend encouraged Abigail to bear the burden as a true patriot: "If your Dearest Friend had not Abilities to Render such important services to his Country, he would not be Called to the self Denying task of leaving for a time His Beloved Wife and Little prattling Brood."[125] Mercy did not provide comfort for her friend but rather pressured Abigail to accept this huge sacrifice. In the end, after major emotional struggle, Abigail consented, and John left for France, taking with him their twelve-year-old son, John Quincy Adams. Deprived of both her husband and her eldest son, Abigail still counted greatly on her friends to understand how much she had personally sacrificed to her sense of public duty.

She also began a small commercial enterprise for which Mercy acted as her agent in selling small items, such as handkerchiefs, tea sets and other luxury items, that John sent from Europe. Mercy was also able to supply her friends with medicinal herbs and thread from the coastal area of Plymouth, and Hannah Winthrop acted as an intermediary between her rural friends and the established artisans of Cambridge. In this way, women friends offered one another concrete material assistance, sometimes even lending money.[126]

In 1778, soon after John and John Quincy's departure, Abigail encouraged her daughter, Abigail Jr., to spend the winter with Mercy in Plymouth. Mercy, the mother of five sons, was delighted to have a teenage girl under her roof, and the girl was apparently happy to be in the company of so genteel a woman. Mercy wrote that she loved "Nabby" more and more the longer they were together. At the same time, Abigail received visits from Mercy's sons and provided one of them with advice on his imminent journey to Europe. Love me, love my children: this mantra, applied earlier to Mme de Sévigné and Mme de La Fayette, is equally suitable for Mercy Otis Warren and Abigail Adams, whose children continued to visit their surrogate mothers for years to come. It also applies to the friendship between Mercy and Hannah Winthrop, as evidenced by the fact that the Warrens' sons lived with the Winthrops when they studied at Harvard.

During the years of John's absence, it was sometimes difficult for Abigail to maintain her friendship with Mercy as before, since John's rise in the world of diplomacy contrasted with James Warren's rockier career. The Warrens became increasingly jealous and bitter. They believed that John Adams and his political allies had betrayed the Revolution in becoming Europeanized—that is, more conservative and more open to hereditary honors. By the time Abigail decided to join John in England, where he had been appointed ambassador to the Court of St. James, the relations between the two families were definitely cooler than they had been in the years leading up to the Revolution. During her time abroad, from 1783 to 1787, Abigail wrote sparingly to Mercy.

Hurt feelings aside, Abigail and John continued to support Mercy's plan to write a history of the American Revolution. To that end Abigail, who had always forwarded copies of her letters from John to Mercy, continued to send Mercy useful information for her work, from the diplomatic centers of London and Paris. Mercy's focus on her historical project probably helped her confront the melancholy state of her family fortunes, which continued to decline. The Warrens never got used to the new republic, with its relatively strong central government, as opposed to the localized agrarian communities they had valued. Mercy continued to rail against what she saw as a drift toward monarchism in the creation of a powerful chief executive.

Ironically, this did not stop her from asking John Adams to support political appointments for her sons and husband. John's blunt refusal and his specific criticism of James Warren, whose "unpopularity" had reached his ears, cut Mercy to the quick. The connection between the Warrens and the Adamses reached a new low, but Abigail continued to correspond with Mercy and paid her a visit prior to John's presidential inauguration in 1797. Abigail assured Mercy: "Old Friends can never be forgotten by me. "[127]

Mercy took her revenge on John Adams in her *History of the Rise, Progress, and Termination of the American Revolution*, published in 1805. She depicted him as a small-minded man of little patience and minimized his contributions as a revolutionary, diplomat, vice president, and president. The book effectively caused a breach in her friendship with both of the Adamses. Yet after Mercy died in 1814, Abigail recognized her friend's unique value: "Take her all in all, we shall not look upon her like again . . . To me she was a friend of more than fifty summers ripening."[128]

Despite the grave differences that separated them in their later years, the Adamses and the Warrens belong together in the story of American history and in the history of women's friendship. During the 1760s and 1770s, when James and Mercy's Plymouth residence was the center of local politics, John Adams had been part of their circle of prominent Massachusetts men and women with republican sentiments. It was at the Warrens' Plymouth home that Abigail first met Mercy and commenced their very long association. As was (and is) often the case with wives and mothers, Abigail and Mercy's friendship privileged their children but ran afoul of their husbands. During the 1780s and 1790s, when John's fortunes rose and James's declined, negative feelings infiltrated the relationship, coming primarily from James and Mercy but also from John. It seems that Abigail made the greatest effort to keep the friendship alive, even if it was at times only "the appearance of Friendship."[129] Abigail's generous, spontaneous spirit contrasted with Mercy's sense of aggrieved self-importance, and, as the balance of power shifted between them, Abigail created a welcoming aura that encompassed many friends, old and new. Still, she never forgot what she owed to Mercy Otis Warren, the older, cultivated woman who had taken Abigail under her wing as a fledging and ultimately helped her to fly on her own.

### Mercy Otis Warren and Catharine Macaulay

Mercy Warren's friendship with the historian Catharine Macaulay began the same year as her friendship with Abigail Adams, in 1773. Since Macaulay lived in England, their relationship was conducted exclusively through letters, except for one visit that Macaulay made to America in 1784 to 1785. In a detailed study of their relationship, historian Kate Davies has written: "Catharine Macaulay and Mercy Otis Warren exchanged letters and ideas with one another for almost twenty years. From cosmopolitan London and Bath and provincial Massachusetts they sustained a close friendship that was almost entirely epistolary and dependent on unreliable transatlantic crossings."[130] Although Warren was three years older than Macaulay, the latter was already recognized in England for her ongoing *History of England*, which would encompass eight volumes when finished in 1783. In contrast, Warren published her verse dramas anonymously until 1790 and would not receive public recognition for her historical works until the end of her life. Nonetheless, from the start they treated each other as intellectual equals and sympathized with each other as personal friends.

Both women initially bonded over their support for the cause of American patriots' demanding greater freedom from imperial rule. In a written address to the British people, Macaulay wrote, "If a civil war commences between Great Britain and her colonies, either the mother country, by one great exertion, may ruin both herself and America, or the Americans, by a lingering contest, will gain an independency." In either case, British subjects would be left only "the bare possession of your foggy islands; and this under the sway of a domestic despot." Such words were not destined to endear Macaulay to her compatriots, but they were sweet to the ears of Mercy Warren and like-minded Americans.[131]

Catharine Macaulay's public statements and personal letters were circulated among Massachusetts republicans of both genders. Mercy's letters to Catharine were more private, though they were nonetheless concerned with politics, which she saw as women's domain as well as men's:

> You see madam, I disregard the opinion that women make but indifferent politicians . . . When the observations are just and do honor to the heart and character, I think it very immaterial whether they flow from a female lip in the soft whispers of private friendship or whether thundered in the senate in the bolder language of the other sex.[132]

"The soft whispers of private friendship" says it all. Mercy, like Abigail Adams, always assumed that women had a role to play in the public realm through the medium of conversation, comparable to that which men had in the Senate. Moreover, since women of that era were kept at a remove from the baser aspects of politics, they might make more rational judgments than men. We have heard this argument before, coming from the English writer Mary Astell earlier in the century.

Instead of apologizing for women's reputedly more emotional nature, Macaulay and Warren turned it to their advantage, even in the political realm. Sharing what they called a language of the heart, they agreed that women's love of family, friends, and country should affect the public discourse. And to emphasize the value of women's feelings and personal relationships, Macaulay asked rhetorically, "Can patriotism dwell in a heart where friendship has no place?"[133] Way ahead of their time, Warren and Macaulay could easily have adopted the twentieth-century feminist slogan that the personal *is* political.

This example of a friendship conducted almost exclusively through letters calls to mind a question that would recur two hundred years later with Internet friendships: can two people really become friends without seeing each other face-to-face? If friendship is marked by an ongoing exchange of feelings and ideas combined with mutual sympathy and respect, then Macaulay and Warren's was indeed a true friendship. Many face-to-face "friendships" never reach similar heights of understanding. Moreover, their commitment to American independence and the formation of a just republic gave them a common cause, which they shared for the rest of their lives.

Their friendship was put to the test in 1784, when Macaulay arrived in the fledgling United States for a visit. She came with her second husband, William Graham, whom she had married in 1778. Since Macaulay had been a widow since 1766, no one begrudged her the right to take a second husband, but her choice—a man only twenty-one to her forty-seven—was bound to create a scandal. Loyal to her friend, Warren defended Macaulay's freedom to marry a man twenty-six years her junior noting the male prerogative to marry a far younger woman with impunity. In this respect Warren was more tolerant than her friend Abigail Adams, who was definitely shocked by the union.

It was not the marriage that caused a temporary breach in Mercy and Catharine's friendship but something quite unexpected, and even slightly ludicrous. During her visit to New England, Catharine chose to frequent a newly established club named Sans Souci, where the Boston elite met for music, dancing, card playing, and gambling. The club was for Mercy Warren an example of the consumer culture and moral decay that had set in to postwar society.

There was more. Warren was accused—wrongly—of having authored a play unflattering to Macaulay, and both were hurt by the discord. The public nature of the Sans Souci quarrel revealed differences between two women who had previously believed they were similar in every way. In this instance, the cosmopolitan British lady and her more provincial

Massachusetts counterpart simply did not see eye to eye on what constituted appropriate social behavior.

Still, they had enough of a backlog of amiable sentiments to patch over the quarrel before Macaulay returned to England. Each treasured the friendship and wanted it to continue; and so they picked up their pens once more and corresponded across the ocean for another six years. Macaulay continued to publish her political thoughts right up until 1790, one year before her death—a death that her transatlantic friend sincerely grieved.

The circle of friends surrounding Mercy Otis Warren was united by the heady days of prerevolutionary discourse and then by the danger and deprivations of the war. Their private friendships were enriched by the sense of meaning that the revolution brought to their lives, not only for those like John Adams and James Warren, who played a public role in the making of a new nation, but also for their wives, who contributed on the domestic front. As wives, mothers, and friends, patriotic women offered their services to the cause they believed in. It was the first of many political causes that would bring American women together for the next 250 years.

### Republican Women in France

The American Revolution officially ended in 1783 with the Treaty of Paris, skillfully negotiated by John Adams, Benjamin Franklin, and John Jay. As an American ally, the French would sign a separate agreement with Great Britain. No one could have imagined in 1783 that the ancient French monarchy would be toppled by its own revolution just a few years later.

During the French Revolution of 1789 to 1795, friendships often ended tragically. Unlike their American counterparts, who generally maintained a modicum of civility even between patriots and British loyalists, French revolutionaries descended into savage class warfare. Many of the friendships formed in the crucible of the Revolution were not destined to last, given the toll in human lives exacted by the guillotine.

France has retained the image of friendship in its famous creed "Liberty, Equality, Fraternity," which is supposedly applicable to women as well as men despite the gendered bias of the word *fraternity*. Though less visible than male friendships, women's bonds with one another also played a role in the French Revolution, one that, after two centuries of neglect, scholars have recently begun to uncover.[134]

The personal friendships of aristocratic women, who were targeted as enemies of the Revolution, ruptured as they went into hiding or, more tragically, mounted the guillotine. Consider the gory end of the Princesse de Lamballe, Queen Marie Antoinette's most intimate friend: her head was paraded past the queen's prison-cell window. Some aristocratic women withdrew to the provinces, where friends were willing to keep them out of sight. Émigré women, who had never worked before, banded together in London to make salable items such as embroidered dresses and painted fans so as to provide for their families. Friendships usually followed class lines, though there were many examples of bourgeois and peasant women showing kindness to threatened aristocrats; some even took over the care of children when a mother had to flee or was imprisoned.

Friendships arose in newly established women's patriotic clubs, such as the Society of Revolutionary Republican Women and regional groups of *citoyennes* (women citizens). Unlike their American counterparts in the 1770s, some French women made their voices heard publicly and collectively in the late 1780s and early 1790s. These women joined forces to write petitions that expressed their grievances and demanded redress from the newly formed government. Working-class women, ranging from fishmongers and flower sellers to artisans and shopworkers, participated in street protests and riots, demanding everything from cheaper bread to the right to bear arms. Though we have few records of the specific friendships made in these female enclaves, it is not hard to imagine the personal relationships that might have developed from their shared efforts to influence the course of the Revolution.

These groups made such a ruckus that, in time, the National Convention outlawed "women's societies and popular clubs" altogether. The all-male Convention decreed in 1793 that women were not capable of exercising political rights and that they should never meet in political associations: "The private functions for which women are destined by their very nature are related to the general order of society; this social order results from the difference between man and woman. Each sex is called to the kind of occupation which is fitting for it."[135] This would be the general attitude of legislators in

France for the next 150 years, right up to 1945, when French women finally got the vote.

### Madame Roland and Sophie Grandchamp

Perhaps the most dramatic record of a patriotic female friendship during the French Revolution is that of Mme Roland and Sophie Grandchamp. Mme Roland became famous right after the Revolution on the basis of her remarkable memoirs, written in 1793 during the five-month stay in prison that preceded her execution. Had she not left behind these memoirs, her life would probably have been reduced historically to her identity as the wife of the minister of the interior, Jean-Marie Roland de la Platière. Certainly no one would have been interested in her friendship with Sophie Grandchamp. Yet it was Sophie Grandchamp who smuggled Mme Roland's notebooks out of prison and attested to her friend's courage on the day of her death.

Let us backtrack. Mme Roland was arrested in connection with Robespierre's 1793 purge of the deputies from the Gironde region, among them Roland de la Platière. While he fled to the southeastern provinces, Mme Roland remained to confront her husband's enemies. Having promptly been incarcerated in her husband's stead, she set out to write what would become the most famous eyewitness chronicle of the Revolution. Mme Roland was singularly situated for such an endeavor, in that she had been privy to the inner workings of revolutionary politics. Moreover, she had already developed a vivid literary style, which she had employed until then only in the service of personal letters and professional documents she wrote for her husband.

During the first two years of their political involvement, the Rolands had been willing to go along with the goals of revolutionaries such as Danton, Marat, and Robespierre, whom they had received regularly in their home. But they were morally outraged by the prison massacres of September 1792, at which point they broke with their former allies and incurred the fate of so many others swallowed up by the Terror.

Sophie Grandchamp was Mme Roland's closest female friend during the last two years of her life. A well-educated, sensitive, bourgeois woman who gave free lessons in astronomy, grammar, and literature, Grandchamp wrote her own recollections in 1806, recounting her close association with Mme Roland, which had begun in February 1791. It was, from her point of view, something of a love story—a mutual rapture of the soul, but one that caused friction between Grandchamp and M. Roland as each vied for Mme Roland's attention and affection. There was even a "lovers' quarrel" between the two women, resolved only when Grandchamp visited Mme Roland in prison.

Grandchamp was the first person to try to understand the psychological reasons for Mme Roland's fervent involvement in revolutionary politics. She believed that her gifted friend had been unhappy with her provincial life before she was able to express herself through her husband's call to national office in 1791. According to Grandchamp's interpretation, Mme Roland had "nourished a secret ambition . . . to appear on a stage where she could deploy all her talents," and once she found that stage, she excelled in communicating her husband's ideas.[136]

As Mme Roland wrote in her own memoirs: "If it was a question of a circular, of a piece of instruction, of an important document . . . I took up the pen which I had more time than he to use."[137] What she took care not to say was that she was in reality the chief force behind the Office of Public Opinion, officially directed by her husband.

In prison, Mme Roland unburdened herself to her one and only female friend. Before she met Grandchamp her friends had been exclusively male, chosen from the republican deputies who had met at her home during her husband's time in office. In that circle she had been something of a queen bee, with little regard for other women, whom she considered intellectually inferior. Sophie Grandchamp was an exception, and she proved to be a true friend in every sense of the word during Mme Roland's last days.

As her execution drew near, Mme Roland asked Sophie Grandchamp to observe her on her way to the guillotine: "Your presence will diminish the fear that this odious journey inspires. I shall at least be sure that a being worthy of me will render homage to the firmness which will not abandon me in such a dreadful ordeal." Sophie followed her friend's instructions. She went out an hour before Mme Roland's departure from the Conciergerie prison and stationed herself at the end of the Pont Neuf, wearing the same clothes she had been wearing when they had last seen each other. As soon as she was able to make out Mme Roland's face in the cart carrying her to the guillotine, Grandchamp stared directly at her: "She was fresh, calm, and smiling . . . Approaching the bridge, her eyes looked for me. I read the satisfaction that she experienced in seeing me at this last, unforgettable rendezvous."[138]

This testimonial of friendship in the face of impending tragedy has added to the legend of Mme Roland as a martyr worthy of her republican principles. Other witnesses testified to the moving words she uttered before the statue of Liberty at the Place de la Révolution (now the Place de la Concorde): "Oh Liberty, how many crimes have been committed in your name!"

Mme Roland never mentioned Sophie Grandchamp in her memoirs, although there is an allusion to one (unnamed) woman who visited her in prison, in addition to three loyal male visitors. Perhaps she didn't mention her because she did not want Grandchamp to be implicated in the fate she anticipated for herself. Or perhaps Sophie Grandchamp just wasn't that important to the story Mme Roland wanted to leave behind. Clearly she was the star in their relationship, with Grandchamp something of a satellite basking in reflected sunlight. Yet Mme Roland chose Grandchamp to be the witness of her final ordeal. She chose a woman who would empathize with her situation and record her tragic destiny. True to her word, Sophie Grandchamp watched without flinching as her dearest friend was carted off to the guillotine, and, true to her ideal of friendship, she subsequently devoted herself to perpetuating Mme Roland's glory.

Late-eighteenth-century female friendships cast in a re-publican mold, whether in America or France, gave many women a new sense of their worth. Although women were denied official rights in both countries, the "soft whispers of private friendship" strengthened their desire to play a greater political role—though that role would not be fully realized until the twentieth century.

# SEVEN

# ROMANTIC FRIENDSHIPS

*"I love you, strongly, openly, intimately. I hold you in high
esteem, I trust in you without reservation."*

—SIBYLLE MERTENS TO ADELE SCHOPENHAUER, 8 MARCH, 1836

*"Why are we to be divided? . . . it must be because we are
in danger of loving each other too well."*

—CHARLOTTE BRONTË TO ELLEN NUSSEY, 20 FEBRUARY, 1837

*"Imagine yourself kissed a dozen times, my darling."*

—MARY HALLOCK (FOOTE) TO HELENA DE KAY (GILDER), LATE 1860S

*"I worshipped Madame Alicia, but that was a tranquil kind of love; I needed a
burning passion. I was fifteen years old."*

—GEORGE SAND, *STORY OF MY LIFE*, 1876

SOMETIME AROUND 1800, FEMALE FRIENDSHIP became romanticized. British, German, French, and American
women—to name only the most visible examples—

began to write each other in the language of lovers. It was not uncommon for girls and women to call each other *darling*, *sweet*, and *precious*, and to speak of loving each other forever. Adolescent girls hugged and kissed and made no secret of the crushes they had on their schoolmates. Women believed that a male–female relationship could never be as "true" as the loving friendship between two women, and they complained bitterly when fiancés and husbands carried away their close friends.

The American historian Nancy Cott calls attention to this new mind-set in her classic book *The Bonds of Womanhood*. She writes that "the diaries and correspondence of New England women suggest that from the late eighteenth through the mid-nineteenth century they invented a newly self-conscious and idealized concept of female friendship."[139] She links this paradigm shift to an increasing identification of women with "the heart"—the symbol of love, compassion, pity, and empathy. Women were expected to open their hearts to their emotionally lacking male counterparts, with the mutual understanding that men were stronger and more rational and thus suited to complement women's extreme sensibility. A collateral consequence of this vision was that all-female friendships were elevated to a higher level of esteem than ever before.

The roots of these changes could be found on the other side of the Atlantic, where the cult of friendship had become almost as popular as the cult of love. The preromantic and romantic movements in literature (roughly 1761 to 1850) encouraged people to feel deeply and to give copious expression to their feelings in torrents of tears, sighs, and poetic utterances. Beginning with Jean-Jacques Rousseau's *Julie, or The New Heloise* (1761) in France and Johann Wolfgang von Goethe's *The Sorrows of Young Werther* (1774) in Germany—novels that headed international bestseller lists—emotion was promoted into a position of cultural regard that rivaled the Enlightenment principle of reason. A woman or man without sensibility, without a "natural" inclination toward affection, tenderness, friendship, and love, was seen as defective. It was in this hothouse atmosphere that romantic friendships took root.

The English romantic writers, who followed in Rousseau's footsteps, also looked to the ancient Greeks for models of friendship. They wanted to emulate all they saw as admirable in Greek friendship, omitting only its homosexual component. The eighteenth-century poet Percy Bysshe Shelley, for instance, who translated Plato's *Symposium*, promoted a vision of men linked in universal brotherhood—an ideal that, despite Shelley's language, did not exclude women. Yet even Shelley would probably have been surprised by the prominence that romantic friendships were to assume among women during the nineteenth century.

Like love affairs, romantic friendships between two girls or two women were often passionate, exclusive, and obsessive. (This won't surprise anyone familiar with middle-school girls today.) Frequently initiated in girls' schools, adolescent crushes could develop into lifelong friendships that continued even after one or both of the women had married. Society not only accepted the sentimental intimacy of two women but actually promoted it as a feminine ideal. While men had their work, clubs, or taverns—depending on their class—what better pastime for a woman than to share the loving company of her best friend?

But if two women took it into their heads to live together without the cover of a husband, mother, or aunt, tongues began to wag. Such was the case for thirty-nine-year-old Eleanor Butler and twenty-three-year-old Sarah Ponsonby, two ladies who left their Irish homes in 1778 with the intention of settling in North Wales. Their first attempt at flight was foiled by their families, as indicated in the following letter, written by one of Miss Butler's relatives: "The Runaways are caught . . . There were no gentlemen concerned, nor does it appear to be anything more than a scheme of Romantic Friendship."[140] In comparison to a heterosexual affair, a "Romantic Friendship" between two women was not the worst thing that could happen to the family name. Eventually, Miss Butler's and Miss Ponsonby's families allowed them to move to Llangollen Vale in Wales, where they lived together for more than fifty years and became famous as the Ladies of Llangollen, in the words of poet Anna Seward, for their "sacred Friendship, permanent as pure."[141]

Butler and Ponsonby's contemporaries did not equate *romantic* with *lesbian*—a word that did not find currency until the late nineteenth century. *Lesbianism* as a term to describe the erotic practices between two women entered the vocabulary of social scientists in the 1870s, as did the word *homosexuality*, used for both men and women (as early as 1869 in the German term *Homosexualität*). Before then, female romantic friendships were generally not thought to be sexual, so a woman could express great affection, and even deep love, for another woman without incurring the suspicion of impropriety.

The American historians of sexuality John D'Emilio and Estelle Freedman caution us against homogenizing complex romantic friendships and turning them all into protolesbian relationships.[142] Some nineteenth-century American women living in partnerships seem to have had little or no interest in sex, whereas other female couples may have been lesbians in today's sense of the word—that is, sexually active. For our purposes, it is useful to think of romantic friendships as stretching across a continuum, from "committed heterosexuality" at one end to "uncompromising homosexuality" at the other, according to the model suggested by historian Carroll Smith-Rosenberg.[143]

### Adele Schopenhauer, Ottilie, and Sibylle

Sentimental attachments between girls and women were already a familiar phenomenon in the early nineteenth century in such countries as England, Germany, and France, where romanticism had captured the imagination of readers and penetrated large portions of popular culture. In the German city of Weimar, the literary circle of men and women surrounding Goethe was the setting in which Adele Schopenhauer (1797–1849), sister of the philosopher Arthur and daughter of the bestselling author Johanna, developed her first great romantic attachment. Her story is one of passionate love for two women, one succeeding the other. Recounted in a recent exhaustive biography, Adele's history exemplifies what was possible for romantic friends within the sophisticated milieu of German intellectuals.[144]

Because we have extensive documentation about the life of Adele Schopenhauer from her letters and diaries, as well as from observations written by her contemporaries, we can present her story here in greater detail than that of other nineteenth-century European or American women whose passionate friendships are sparsely documented. Adele's friendships, as with all the examples selected for this chapter, must be understood in relation to their specific cultural setting and historic moment. The contemporaneous English writer Anna Jameson saw Adele's partnership with Sibylle Mertens as "essentially German." She would not have expected to find the same characteristics in an English female couple.[145]

As a child Adele Schopenhauer came to Weimar with her widowed mother, who quickly established a successful literary salon (the French word had been adopted by the Germans) that attracted even the great Goethe. He befriended not only Johanna but also her precocious daughter, Adele, and took such a liking to her that he oversaw her cultural education and remained her avuncular friend until the end of his life.

Adele's childhood friend Ottilie von Pogwisch was her beloved companion until she was twenty-six. In one of the many letters that have survived from their lifelong correspondence, Adele wrote to Ottilie: "With all the strength of my being, I love you . . . I cannot live if you are unhappy, because you alone know all my thoughts and experience everything as I do."[146] Here friendship is presented as total identification with the other, recalling Aristotle's proverbial one soul in two bodies.

Ottilie returned Adele's ardor, as can be seen from her letter of July 1816, in which she wrote: "I'll be eternally happy as long as you are mine—and won't you remain mine always? What could separate us?"[147] Ironically, what separated them was Ottilie's decision to marry August von Goethe, the writer Goethe's only surviving son. Nevertheless, Adele remained Ottilie's closest confidante throughout what proved to be an unhappy marriage; she even facilitated her married friend's affairs with other men. But Adele complained to her diary that they saw each other less frequently than before, and that no one would ever love her as deeply as she loved Ottilie.[148]

In 1828 Adele changed her mind when she met Sibylle Mertens, a married woman her own age and a mother of six. It did not take Adele long to fall under Sibylle's spell, as evidenced in a letter she wrote to Ottilie: "Other than you, I believe I have never loved anyone so much."[149]

For the first four years of Adele's relationship with Sibylle, Adele and her mother would spend the summer with Sibylle at her country home in Unkel, situated some thirteen miles from Bonn. There, as she did in Cologne during the winter, Sibylle ran a salon that was similar to Johanna Schopenhauer's in Weimar. Sibylle's interest in archaeology and antiquities added a special dimension to her circle of writers, philosophers, musicians, and artists—a circle in which Adele felt very much at home. In her correspondence with Ottilie, Adele enumerated Sibylle's many talents: she read mythology, history, and Latin authors in translation; played the piano with great skill; and was interested in art, sculpture, and poetry. Moreover, since Sibylle's unsympathetic husband, Louis Mertens, remained at the couple's primary residence in Cologne attending to his work as a banker, Adele and Sibylle were able to spend whole days together unhindered by an intrusive male presence.

The second summer of their friendship was even more idyllic than the first. Reunited in Unkel, the two women became "indispensable" to each other. Adele wrote to Ottilie words that might have startled her former best friend: "I cannot remember such a trusting friendship in my life . . . I think it's best for you to compare us to a pair of people who find each other late [in life] and then marry. If she were to die—I would jump into the Rhine."[150]

Together the women read Goethe's *Wilhelm Meister*, wrote poems, and had the pleasure of receiving the first issue of a weekly newspaper called *Chaos*, which Adele had begun in Weimar with Goethe's backing. Like many nineteenth-century women on both sides of the Atlantic, they even slept together in a double bed, referred to in German as a "French bed." For about two months, Sibylle managed to remain in Unkel without her husband and five of her six children, the ostensible reason being her poor health.

After the first four years of their loving friendship, Adele began to have doubts as to whether she was the sole possessor of Sibylle's heart. When the famous poet Annette von Droste-Hülshoff appeared on the scene, Adele became bitterly jealous. She found even more reason for jealousy later, when Annette was Sibylle's caretaker during a long illness.

Another potential rival was Anna Jameson, who joined their circle through her friendship with Ottilie von Goethe. In one of her books, Anna Jameson portrayed the Adele/Sibylle couple, without naming them, in the following manner:

Opposed to each other in the constitution of their minds . . . yet mutually appreciating each other: both were distinguished by talents of the highest order and by great originality of character, and both were German, and very essentially German: English society and English education would never have produced two such women.[151]

Anna might have been referring to their serious, non-playful, highly intellectual natures, but she might also have been hinting at Sibylle's domineering character. Annette von Droste-Hülshoff, in her own picture of Sibylle, described her as a "capable house commander" who ranked among the women "who wear the pants."[152] Others saw Sibylle mainly as eccentric, but all agreed that she had a strong intellect. At a time when few women ventured far from the three German K's—*Kinder, Kirche, und Küche* (children, church, and kitchen)— Sibylle Mertens had an interest in archaeology that would lead to excavations in Rome and a fine collection of antiquities, some of which eventually found their way to the British

Museum. Sibylle Mertens was truly a "Renaissance man" in terms of her many accomplishments.

Her stay in Italy with some of her children from 1835 to 1836, again ostensibly for reasons of health, initiated a long separation from Adele. Though she had another woman companion, she expressed longing for Adele in her diary and sent her loving words: "It is a life-need for me in this distant place to say to you: I love you, strongly, openly, intimately. I hold you in high esteem, I trust in you without reservation."[153]

Adele and Sibylle were not to be truly reunited until 1842, when Louis Mertens's death permitted them to be together continually. By then Adele had become better known as an author of stories and novels—though she would never be as popular as her mother, not to mention her philosopher brother, whose great fame would come posthumously. Sibylle's circumstances, too, had improved due to a substantial inheritance from her husband. Now she was free to winter in her beloved Italy, with Adele at her side. They settled into a routine of visiting churches, palaces, and museums, which allowed them both to blossom. Adele would later write to her brother, Arthur, that living in Italy with Sibylle in 1844 loosened her up and allowed her to pursue new interests and ideas.

Sibylle rented a small house in the middle of Rome for the 1845 to 1846 winter season. Adele took two rooms nearby, though for the most part they lived together. Sibylle established an international salon, which boasted the presence of "salon ladies, learned men, clergymen, good housekeepers, artists, musical celebrities, tourists, *Monsignore* and women writers, the businessman and the prince, the ambassador and the doctor of law, the doctor and the elegant lady," according to Fanny Lewald, a German writer familiar with the gatherings.[154]

Even Ottilie von Goethe came and lent her famous name to the brilliant "Tuesdays" organized by Sibylle Mertens. More important for Ottilie than the salons was the comfort of her two friends after the loss of her daughter, who had been stricken down by typhus in Vienna, where Ottilie had relocated. Ottilie also benefited from the emotional support of Anna Jameson, a close friend since their meeting a dozen years earlier. Anna had stayed in Vienna for two months to help Ottilie bear her grief, and soon after her departure Anna sent Ottilie a loving message: "When I parted from you, dearest Ottilie, I felt more deeply than I had ever felt before how much I really love you."[155]

These words are representative of the rhetoric that English and German women were wont to use in their intimate friendships. "I love you" and "ich liebe Dich" dance off the page in a hundred different formulations, as women conveyed to one another their innermost emotions. Should we see them as indications of what we would today call lesbian relationships? Does it matter? Obviously one can feel loving emotions for members of one's own gender that are not necessarily sexual. What is certain is that Anna Jameson, like Adele Schopenhauer and Sibylle Mertens, came to the aid of her grieving friend when she needed her. Romantic diction apart, this is what women did and still do for their friends.

The last period of Adele's life—she was to die in 1849—was spent largely in Italy with Sibylle during the winter season, and in Germany for the rest of the year. These years were marked by complicated legal dealings with Sibylle's children, resentful witnesses of their mother's close attachment to Adele. Adele, for her part, was also involved in financial transactions with her brother, who, though he kept his distance from his sister as he had from their mother, managed to get the better part of their inheritance. Nothing good can be said about Arthur's relationships with women (or with men, for that matter). The noted misogynist made his contempt for women explicit in his writing, most notably in *Parerga and Paralipomena*. Unlike Adele's women friends, he did not even pay her a visit on her deathbed.

Adele had the misfortune of being the sister of such a misogynistic man, but she also had the good fortune of knowing Goethe for the first thirty-five years of her life, and Goethe was by no means hostile to women. On the contrary, he loved them in many ways, and he was broad-minded enough to understand Adele's love for other women. In private, Germans might call such women "unnatural" or "eccentric," but they did not exclude them from polite society, especially when the women had intellectual and financial capital.

Adele spent her dying days in Sibylle's house in Bonn. Each day Sibylle placed roses next to her bed, and when the pain became too severe, she gave her alcohol and opium. Shortly before she died, Adele was gladdened by a visit from her first love, Ottilie, now linked in a friendship with Sibylle that would continue after Adele's death. Adele wrote her will on August 10, 1849, leaving to Sibylle all her earthly possessions, excluding a large sum of money designated for Arthur. On August 25, she died in the arms of her beloved partner.

Sibylle washed Adele's body and placed it in a coffin, which was then interred in the old Bonn cemetery. She had the tombstone inscribed in Italian with a text that read: "Here lies Luise Adelaide Lavinia Schopenhauer, after a life of fifty-two years, outstanding in heart, spirit, talent, the best of daughters, and true to her friends . . . This monument was erected by her inconsolable friend, Sibylle Mertens-Schaaffhausen." It is one of the rare testimonials to female friendship inscribed—in any language—on a tombstone.

### George Sand

The lives of literary figures are, of course, better documented than those of other mortals. In her autobiography, the celebrated French writer George Sand (1804–76) wrote expansively about the girls who shared her life in a Parisian convent school when she was a teenager, known to her classmates by her given name, Aurore Dupin.[156] Aurore's closest friends were three girls named Sophie, Fannelly, and Anna, all bound by a code that required them to draw up a list of their best friends in an immutable order based on chronology, despite changes of the heart: "Once we had given a girl first place, we did not have the right to take it away in order to give it to someone else. The rule of seniority was law."[157]

Her friend Fannelly de Brisac was only number three on the list, but number one in her heart. Sand was to write that "despite the list, despite the rule of seniority, despite exchanged promises, I could not keep myself from feeling that I loved Fannelly more than all the others." She described Fannelly as a blue-eyed angelic creature, "small, blond, as fresh as a rose, and with an expression so lively, so open, so kind, that it was a pleasure to look at her." Even though they did not see each other after they left the convent (which was the case of countless girls in Europe and America after their departure from boarding school), Sand had no doubt about the permanence of their feelings: "There is one thing I am as sure of as I am of my own existence, and that is that Fannelly still loves me warmly and tenderly, that no cloud has obscured the irresistible and total understanding we felt for each other thirty years ago, that she never thinks of me without knowing she loves me and that I love her in return."[158] Sand was one of the first thinkers to identify adolescence as a distinct age characterized by sincere affections that remain dear to one's memory precisely because they were so spontaneous and innocent.

In retrospect, she critiqued the convent for its unnecessary exaggeration of chastity and fear of intimate friendships: "We were forbidden to walk about in twos; there had to be at least three of us together at a time; we were forbidden to kiss; there was concern over our innocent letter-writing; and all this would have given us something to think about if we had had so much as an inkling of the evil instincts that were apparently attributed to us."[159] The early Church warnings against "particular friendships" had embedded themselves permanently within the convent mentality.

As an adult, Sand was subject to rumors about the nature of her relationship with the actress Marie Dorval, with whom she began a loving friendship when she was twenty-nine and Marie thirty-five. Their correspondence and Sand's autobiographical account project the image of two ardent souls drawn together by the sheer force of their magnetic personalities. Each found in the other a complementary artistic temperament and a heart sensitive to the most intimate feminine confidences. Of course, George Sand is much better known for her romantic affairs with men—the poet Alfred de Musset and the composer Frédéric Chopin, to mention only the most prominent of her lovers.

But long before she became the famous author George Sand, Aurore Dupin had followed the course of many girls raised in the provinces and was allowed to play freely with both girls and boys. When sent to Paris at the age of thirteen and enclosed within an all-female world, she directed her affections to the girls around her, and to select nuns. Her worshipful love for Sister Alicia "shone alone, above them all," like the sun.[160] Crushes on girls their own age and on older mother figures like Sister Alicia were the norm for girls segregated by sex in convent schools. When compulsory public education was introduced in France in the 1880s, coeducation was allowed for younger boys and girls only if the municipality could not afford two elementary schools, while all high schools (lycées) were single-sex, structured intentionally to keep girls and boys apart. It is not surprising that girls who were confined to same-sex company showered upon one another the explosive emotions unleashed by puberty, as they still do today.

Later in life, after Aurore Dupin/George Sand had married, become a mother, returned to Paris to write, and separated from her husband, she had many close friends of both sexes, including the writer Gustave Flaubert. This scenario was undoubtedly unusual. Marriage and family occupied most French women full-time, as they did most women throughout the world. If they had the leisure for friends, these would be found first among their kin—sisters, cousins, and aunts—and then from women living in the same village, town, or city. Proximity still counted in the nineteenth century. But so did girlhood friendships. Those one had known in childhood and adolescence, those who had received one's awkward confidences, spontaneous embraces, and protestations of eternal loyalty—those would remain in a woman's heart, often for a lifetime.

### Charlotte Brontë

Intense, loving friendships were similarly fashionable in Anglo-American schools as early as the 1820s and 1830s, when romanticism was at its zenith. The famous English writer Charlotte Brontë (1816–55) is known to have established a romantic friendship with Ellen Nussey when they were together at boarding school. Later, when Charlotte was twenty-one and working as an assistant teacher in the school they had both attended, she wrote to Ellen these deeply affectionate words:

"Why are we to be divided? Surely, Ellen, it must be because we are in danger of loving each other too well—of losing sight of the *Creator* in idolatry of the *creature*."[161] The daughter of an Anglican clergyman, Charlotte was less concerned with the gender of the person she loved than she was with the fear of slighting her devotion to God by loving another mortal too intensely. Their lifelong emotional attachment was mediated primarily through letters, right up until Charlotte's premature death from complications of pregnancy.

### Romantic Poetry Between Women

Anglo-American culture in the nineteenth century encouraged women to write and publish soulful poetry about their same-sex friends, following the lead of Katherine Philips two centuries earlier. These poems were no less passionate than heterosexual lyrics in their expressions of tenderness, longing, and eternal commitment. They often contained fond recollections of the past and fantasies of future meetings, as well as occasional fits of jealousy or prolonged grief at a loved one's death.

Dorothy Wordsworth (sister of the English poet William Wordsworth) wrote a poem to her friend Julia in 1827: "Contented to lay bare my heart / To one dear Friend, who had her part / In all the love and all the care / And every joy that harboured there." This idyllic friendship was destined to end, as many did and still do, when Julia married: "We parted, sorrowful; by duty led; / My Friend, ere long a happy Wife." Dorothy, unmarried, presented her sweet memories as testimony to her ongoing connection with her distant friend's beating heart:

> *Thou dost not ask, thou dost not need*
> *A verse from me; nor wilt thou heed*
> *A greeting masked in laboured rhyme*
> *From one whose heart has still kept time*
> *With every pulse of thine.*[162]

Other nineteenth-century English writers, including Elizabeth Barrett Browning and Christina Rossetti, continued to promote the cult of romantic friendship throughout the reign of Queen Victoria, from 1837 to 1901. To cite but one example, here are two stanzas from Christina Rossetti's melancholy poem "Gone Before," written after the death of her close friend:

> *She was most like a rose, when it flushes rarest;*
> *She was most like a lily, when it blows fairest;*
> *She was most like a violet, sweetest on the bank:*
> *Now she's only like the snow cold and blank.*
> . . . . . . . . . . . . . . . . . . .
> *Earth is not good enough for you, my sweet, my sweetest;*
> *Life on earth seemed long to you tho' to me fleetest.*
> *I would not wish you back if a wish would do:*
> *Only love I long for heaven with you.*[163]

The subject of death as the final separation between friends and the Christian hope for reunion in heaven run through this and many other Victorian poems of Rossetti's generation.

American poets also picked up the theme of female friendship, expressing it in texts of varying artistic worth. Here is a typical example, from "The Garden of Friendship," written by Frances Osgood in 1850:

> *I'm weeding my garden of Friendship,*
> *Till only its flowers remain.*
> . . . . . . . . . . . . .
> *And you, in your delicate bloom, love,*
> *Pure, tender, and graceful and true,*
> *Shall be the queen-rose of my garden,*
> *And live on Love's sunshine and dew.*[164]

*Godey's Lady's Book*, by far the most popular American journal for women in the nineteenth century, extolled sentimental bonds between women in many of its articles, stories, and poems, the latter meant for young women to copy into their friends' autograph albums and stitch into friendship quilts. Society as a whole embraced the belief that girls and women could and should love one another loyally—at least until a proper husband came along.

### American Schoolgirls

As in England, American boarding schools, stretching from north to south and east to west, were natural breeding grounds for intense female friendships. Sent away from their homes during their teen years, middle- and upper-class girls were encouraged to bond closely with their school friends and teachers. In her book *Scarlett's Sisters*, Anya Jabour has documented the widespread prevalence of romantic friendships between schoolgirls in the antebellum South. She presents a convincing panorama of Southern girls who looked to one another for emotional fulfillment during that privileged time of life when they were no longer under the direct rule of their parents and not yet subject to the relentless demands of wifehood and motherhood. Their letters, diaries, memoirs, and autograph books are drenched with longing for "ardent love" and "a single true friend."[165]

Loula Kendall from Alabama gave vent to intense feelings for her "darling," "precious" Susie: "I love her so passionately." When they were apart, Loula complained of missing the feel of her friend's "sweet ruby lips" pressed to her own. When Susie left school for good, they promised to "think of each other every evening at twilight."[166]

As in the case of George Sand, the Southern girls had their own prized rituals surrounding their attachments. It was understood that girls could share with their best friend a classroom desk, dormitory alcove or room, or, most intimately, their bed. Some girls exchanged flowers or candy, or more permanent items, such as locks of hair or rings. Others wrote poetry or sang sentimental songs in honor of their loved one. They hugged and kissed publicly and held hands for all the world to see. All wept when they parted at the end of term and sobbed even more when their school years came to a close.

Their autograph albums, exchanged when they left school around the age of seventeen, attest to the primary place friendship played in their youthful lives and express the hope that it would continue, even at a distance. In this vein, in 1859, a classmate from the Greensboro Female Academy in North Carolina copied lines from a well-known hymn into the autograph book belonging to Martha Ann Kirkpatrick:

*Blest be the dear uniting love*
*That will not let us part!*
*Our bodies may far off remove,*
*Yet still I hope we're joined in heart.*[167]

Autograph books were a staple of American boarding and public schools well into the twentieth century.

When Lucy Catherine Moore Capehart looked back upon her time at St. Mary's School in North Carolina during the 1850s, she reminisced nostalgically about a form of friendship that seemed to have disappeared: "I know not if it is the custom now [1906] for school girls to have sweethearts among their own sex, but it was in those days; such devotion you cannot imagine . . . My sweetheart was Ellen Brent Pearson; to get a smile from her . . . made me supremely happy."[168]

Lucy Capehart's comment reminds us once again that even something we consider as eternal as friendship takes different forms in different times. In her own lifetime, she had seen a dramatic change in the practice of friendship among girls. By 1906, it was no longer "fashionable" for schoolgirls to have passionate attachments to members of their own sex—attachments that middle- and upper-class American girls had enjoyed without censure during Lucy Capehart's youth.

### Mrs. Luella Case and Miss Sarah Edgarton

After the ubiquitous friendships experienced in girlhood, American women sometimes formed same-sex attachments in adulthood that rivaled the feelings they were supposed to have for men. Such was the case of two Massachusetts women, Mrs. Luella Case and Miss Sarah Edgarton, aspiring authors in the 1830s and 1840s. As interpreted by two knowledgeable scholars, their close companionship was "expressed in the amorous language which was the characteristic style of feminine friendship in the nineteenth century."[169]

Like many American women of this era, Mrs. Case and Miss Edgarton first met at church. They both attended Universalist conventions and wrote for Universalist publications. Before long, they were writing to each other soulful letters that expressed a shared sensibility and a longing to spend time together. Mrs. Case opined philosophically: "Life is short, and kindred spirits are few . . . and poor human nature has so many jarring strings, that, after all, friendship is more to be worshipped as an ideal good, than a real, and possible thing."[170] Miss Edgarton fantasized poetically: "Come to me when the flowers and birds are come, and we will dwell with them in greenwood bowers—and our papers and books shall be with us and we will read and talk, and form plans and be the happiest editorial wood-nymphs that ever watched over the flowers."[171] Mrs. Case picked up the bucolic fantasy, adding a swipe at the hornet—ostensibly her minister husband—who would need to be shoed away: "If we could live and work together . . . while you are kissing away the venom of some angry hornet from my lips, why, then, I should be ever very happy."[172]

The two women maintained their sentimental friendship for five years, until 1844, when Sarah Edgarton became engaged to a divinity student. In 1846 she married the newly ordained pastor, whereas Mrs. Case left her minister husband (for reasons unknown); she lived without him until her death ten years later. Sarah Edgarton's life was cut short unexpectedly: in 1848, a year after the birth of a daughter, she died abruptly, leaving behind a presumably saddened husband and the woman friend with whom she had shared her early pastoral dreams.

### "Boston Marriages"

When the relationship between two women was platonic, like this one presumably was, it could have all the trappings of heterosexual love, including social acceptance. However, homosexual relations—considered here as a variant of romantic friendship—could not be conducted openly among nineteenth-century American and English women. It is true that sometimes they were able to live together in unions that were presumed to be "innocent"—that is, nonsexual. In America, such unions came to be known as "Boston marriages."

Boston marriages allowed two single women, usually working women from the middle class, to share their lodgings and
their lives. According to Lillian Faderman, who has studied
such unions extensively, these relationships "were probably
most often not genital in their expression," though they were
certainly "intensely passionate."[173] Society accepted Boston
marriages as an economic alternative to legal marriage precisely because they did not appear to be sexual and, therefore,
did not constitute a threat to heterosexuality.

Given the secrecy that female sexual relations were required to maintain in the past, the historian is hard-pressed to
find letters, diaries, memoirs, or poems that document what
Smith-Rosenberg has called "uncompromising homosexuality" among women. And yet, such are the surprises of scholarly
research that one sometimes comes across totally unexpected
materials to fill in the blank spaces. This is what happened
with the discovery of the Anne Lister diaries.

*Anne Lister, Marianna Belcombe Lawton, and Ann Walker*
The diaries of Anne Lister, written between 1806 and 1840,
contain four million words, some of which were written in
code. These journals remained intact within Anne's family
home, Shibden Hall in Halifax, England, until 1887, when
one of her distant relatives, John Lister, decided to publish
extracts from them. With the help of an antiquarian friend,
John Lister was able to break the coded sections, but he was so
shocked by what he found that he never published a word. He
replaced the journals in Shibden Hall, where they remained,

presumably untouched, until 1934, when the estate was given to the people of Halifax and turned into a museum. At that time the town clerk contacted the antiquarian, who reluctantly produced the key to the code. Even with this key available to researchers, the secrets contained in the journals were concealed for another half century. Then, in the early 1980s, a local scholar, Helena Whitbread, began to work on the diaries and eventually produced two versions for the general public, which ultimately inspired a BBC television drama.[174]

Without access to the coded passages—about one-sixth of the diaries—one gets a limited picture of Anne Lister's life as a respectable, if somewhat odd, member of Halifax's small-scale landed gentry. Her "oddity," by her own admission, consisted of preferring the company of women and vowing never to marry, as well as wearing only black clothes. Her preference for women is documented in her journals as early as 1806, when she was in boarding school and initiated the first of her romantic friendships. Thereafter she enjoyed several intimate relationships with women, including the great love of her life, Marianna Belcombe Lawton, which continued even after Marianna's marriage. Any attempt to construe Anne Lister's friendships as platonic is exploded when one reads the passages in the code composed of Greek letters and algebra that Anne invented as a teenager. These passages are explicit in detailing the sexual nature of her relationships with women: what she called her "amorosos" often consisted of genital contact leading to orgasm, indicated by the word *kiss* or an *X*. Although Anne was to have many sexual adventures during the course of her life (1791–1840), her liaison with Marianna Belcombe Lawton seems to have been the most profound. She began to love Marianna in 1812, when they were both single, and they continued to love each other, despite the obstacles incurred by Marianna's marriage. Ten years into their romantic friendship, Anne penned this coded entry in her diary at Shibden Hall: "M—very low tonight. We sat up talking & consoling each other & latterly in playful dalliance & gentle excitement. Our hearts are mutually and entirely attached. We never loved & trusted each other so well & have promised ourselves to be together in six years from this time. Heaven grant it may be so."[175] The hope for a life together in six years was based on the fact that Marianna's husband, Charles Lawton, was considerably older than his wife and was apt to die before her.

When Anne Lister inherited Shibden Hall upon the death of her uncle in 1826, Marianna temporarily left Charles. The diary records the two women's reunion in graphic detail: "Slept very little last night. Talked almost the whole time till about 4 in the morning. Went to Marianna four times, the last time just before getting up. She had eight kisses and I counted ten."[176] The two women were apparently happy together, and sexually very compatible; but, perhaps fearing the scandal that would ensue if she left her husband for good, Marianna resigned herself to going back to Charles.

Anne was still determined to find someone to share her life as a companion and de facto wife. This time she was more practical, and more successful. She latched on to a slightly younger, somewhat unstable woman named Ann Walker, the wealthy heiress of a neighboring estate. Though their relationship lacked the emotional intensity Anne Lister had known with Marianna, it was not lacking in sexual pleasure. Witness Anne's diary entry of January 8, 1834: "Goodish touching and pressing last night—she much and long on the <u>amoroso</u> and I had as much kiss as possible with drawers on." And again on February 10, 1834: "She was at first tired and sleepy but by and by roused up & during a long grubbling said often we had never done it so well before." Two days later, the couple decided to give each other rings in token of their union. On February 27, Anne Lister was confident enough to write that their union was "now understood to be confirmed for ever." By May, she was totally satisfied that Ann Walker had become fully attached to her: "She says she gets fonder & fonder of me and certainly seems to care enough for me now. I think we shall get on very well."[177]

During the summer of 1834, using their combined incomes, the two women traveled in style to France and Switzerland. Upon their return, they settled into Shibden Hall and lived together like a married couple, with the additional presence of Anne Lister's aunt. Though there was some scurrilous talk among relatives and neighbors concerning the union of the neurotic young heiress with the forceful woman sometimes referred to as "Gentleman Jack," the locals gradually came to accept their arrangement. Their not-so-clandestine marriage survived until Anne Lister's premature death during a trip to Russia in 1840. It was left to the shy wife to bring her partner's body back for burial in the Halifax Parish Church.

In the afterword to her edited volume of the Lister diaries, Jill Liddington has made this appraisal of the two women's relationship: "Anne Lister did certainly take advantage of Ann Walker's wealth and of her loneliness: it remains hard to read the 1833–36 diaries presented here without seeing their marriage as strength manipulating weakness."[178] One wonders how this differed from many other Victorian marriages between a man and a woman. Clearly Anne Lister, as the "husband," was concerned with the financial management of their estates, and she does seem to have put money matters above all other considerations, including those of a romantic nature. Yet she did care for Ann Walker, who, given her pliable nature, seems to have adapted with grace to their unusual arrangement.

Anne Lister's long-term lesbian romances, first with Marianna and later with Ann, fall at one end of the romantic friendship continuum, where love intermingles with sex. Whatever term we use, Anne Lister's ability to find fulfillment in her personal life and live with a considerable measure of respectability speaks to her determined character and the civility of her peers. The English, for all their sense of propriety, seem to have a soft spot for eccentrics. On the whole, they tolerated Anne Lister's self-proclaimed "oddity" and allowed her to maintain a façade of platonic friendship, regardless of the rumors suggesting that the two women did not deny themselves the carnal pleasures of heterosexual couples. Victorian society preferred to see women as asexual creatures, angels in the house, devoid of physical desire.

Before the late nineteenth century, women in England, Europe, and America were able to express their love for one another publicly without causing too many raised eyebrows. Though sexual contact between women was taboo and had to be hidden, many romantic friendships were openly marked by such physical contact as hugs, kisses, or cuddling in bed, especially among adolescent girls but sometimes among mature women as well. Thus Eliza Schlatter wrote from New Jersey to Sophie DuPont in Delaware after the latter's marriage: "I wish I could be with you present in the body as well as the mind & heart—I would turn your *good husband out of bed*—and snuggle into you and we would have a long talk like old times."[179]

### Mary Hallock Foote and Helena de Kay Gilder

The same sensual longing is evident in letters written by the young American woman Mary Hallock Foote to her intimate friend Helena de Kay Gilder. Mary and Helena met in New York City in the 1860s, when they were both studying art at the Cooper Union. They shared the excitement of being young together in the big city and developed an indissoluble friendship that was to last for half a century. During this time they constantly exchanged letters, about five hundred of which have been preserved—four hundred of Mary's and one hundred of Helena's.[180] These documents reveal how two women, coming from very different backgrounds and ultimately separated by an entire continent, preserved a sense of intimacy and unbroken affection that defied time and place.

Mary came from old Quaker stock that had cultivated land in Milton, New York, for five generations. Her family was comfortably middle class, but quietly modest and given to rural pleasures. Helena, born into the upper crust of New York society, had spent much of her childhood in Europe. She would marry a gentleman poet-publisher, Richard Gilder, and come to know some of the most influential people of her day, including President Grover Cleveland and his wife. Mary would marry an engineer who took her to mining camps in the West, where their lives were always financially precarious.

In the letters Mary wrote from Milton, New York, during the 1860s and early 1870s, when Helena was seventy-five miles away on the other side of the Hudson, Mary comes across as the "suitor." She addresses her adored friend as "My beloved Helena," "My dear Girl," "My dearest Girl," "My own dear Girl," "My darling," and "My darling Girl," and signs her letters with such expressions as "Your attached friend," "Yours most lovingly," "Always lovingly thine," and "Will you love me always?" Many of these letters pulsate with Mary's longing to be with Helena and her resentment at the "fates" that kept them apart. For almost a decade, her love seems to have been reciprocated.

Yet Mary was ultimately forced to recognize that her hope for a permanent union with Helena would not be realized. She conceded, nostalgically, that their self-sufficient twosome was a thing of the past. "I shall always remember . . . that for a time at least, I fancy for quite a long time, we might be sufficient for each other."[181]

Even after Helena was married and became a mother, Mary wrote these words that fully expressed her heartfelt longing:

*Your letter came this morning just after Phil & I had been getting your room ready and making your bed—our bed where I thought I should lie tonight with my dear girl's arm under my head—It gave me a queer little sick trembly feeling that I've had only once or twice in my life—and then I thought I must see you—not to "talk things over"—I don't care about things—I only want you to love me . . . I wanted so to put my arms around my girl of all the girls in the world and tell her that whether I go to N. Y. or stay home, whether she signs herself "very truly your friend" or "your dearest of girls," I know I love her as wives do (not) love their husbands, as <u>friends</u> who have taken each other for life— and believe in her as I believe in my God.[182]*

Surely Mary's feelings for Helena represent the kind of ultimate commitment that is more commonly associated with marriage or religious faith than with friendship, as Mary herself openly stated. Though few friendships in any age reach these heights, many women of Mary and Helena's generation had same-sex attachments that were deeply emotional and sensual without necessarily being sexual. They went to bed with a head on a friend's bosom, applied oil to each other's bodies during pregnancy, assisted each other in childbirth, and even slept with a dying friend until her last breath. Despite the potential conflict between female friendship and marriage, it was not uncommon for a bride to bring her sister or best friend along on her wedding trip so as to help her adjust to the physical and emotional demands of married life.

These women were not subjected to the harsh censure similar women would incur in the 1880s and 1890s, when social scientists began to label same-sex love as pathological. Sigmund Freud would say they had failed to develop into "normal" male-loving adult women, and Richard von Krafft-Ebing would excuse them on the grounds that they had been born with a congenital defect. Newly minted technical words like *homosexual*, *invert*, and *lesbian*, all with negative connotations, slowly slipped from pseudoscientific vocabulary into public awareness and contributed to the gradual demise of unself-conscious romantic friendships between women.

*Unself-conscious* is the operative word here, for surely after 1900 romantic friendships continued between women—only now they are called *lesbian*, with an emphasis on erotic feelings and sexual practices.[183] In this book we have used the word *lesbian* regarding women's friendships in the past if there was concrete evidence of sexual activity, as in the case of Anne Lister. Further, we do not assume that women who lived together in ostensibly platonic relationships, such as the Ladies of Llangollen (Eleanor Butler and Sarah Ponsonby), were suffering from repressed sexuality, to use the Freudian terminology. Women's intense attachments to each other cannot be rendered into a word or a formula, unless it be a paraphrase of Montaigne's famous explanation of why he loved La Boétie: because it was she, because it was me.

In the romantic poet William Wordsworth's sonnet about the Ladies of Llangollen, he called them "sisters in love," an epithet fitting for his age and even for ours.[184] *Sisters* implies a deep, indissoluble union, and, under totally different circumstances, the word would figure prominently in the 1960s feminist rallying cry "Sisterhood is powerful." As for *love*, the second term in Wordsworth's phrase, it will always hold a mystery, whether between two women, two men, a man and a woman, or between people who label themselves straight, gay, lesbian, bisexual, transsexual, transgender, queer, or a combination thereof. In the future, there will undoubtedly be more terms for gender variance and more compound expressions for what Butler and Ponsonby's contemporaries called "romantic friendship."

# EIGHT

# QUILT, PRAY, CLUB

*Reader, did you ever go*
*Where the ladies meet to sew,—*
*Needle, thimble, thread in hand,*
*Old and young, a happy band?*
*Take a seat and hear the chat,*
*Now of this and then of that—*
*Shoes or sofas, songs or bread,*
*Books or dresses, lace or thread*
*The last wedding, and the bride,*
*And a little world beside.*

—"THE SEWING CIRCLE," 1852

*"Let us encourage social life, for it is the centre, the heart, of the club; upon it the very*
*existence of the club depends, for unless a club is social it is not co-operative, and un-*
*less it is thoroughly co-operative it is not successful."*

—CLUB MEMBER, ASSOCIATION OF WORKING GIRLS' SOCIETIES CONVENTION, 1890

*"[Our club] has represented the closest companionship, the dearest friendships, the*
*most serious aspirations of my womanhood."*

—JANE CUNNINGHAM CROLY, 1899

WOMEN TEND TO GATHER. WHATEVER they call themselves—circle, club, society, association, gang, bunch—their groups have a charm and a power stronger than the sum of their parts. Not everybody in a group is "best friends" with every other member, and yet all the different relationships within the group enrich and enliven the one-on-one friendships within it. Such bonds have given American women a sense of community and, often, agency that has rarely been available to their gender.

In the New World, women's groups became cornerstones of American society. Originally formed out of necessity, these groups lent badly needed cohesion to the new nation, whose inhabitants were spread far and wide, separated by dirt roads that were impassable in snowy and muddy conditions. While Americans took pride in their independence, self-reliance, and doggedness, some jobs, such as barn raisings and harvests, simply took more than two hands to accomplish. When women gathered to tackle communal labor, especially in groups that met time and again, those groups took on a sort of umbrella personality, and its members developed feelings of friendship not only for one another but also for the group itself.

### Women's Work

American women in the early 1800s, as in every previous era, formed and maintained friendships during the daily activities that held body and soul together. Women's routines consisted in the proverbial "never-done" work of providing food, clothing, and shelter for their families. They were also expected to bear and raise a brood of future responsible citizens for the fledgling republic.

Sewing was on every woman's to-do list, usually from the time she was old enough to hold a needle. Edith White remembered of her 1860s childhood, "Before I was five years old I had pieced one side of a quilt, sitting at [Mother's] knee half an hour a day, and you may be sure she insisted on tiny stitches."[185] Girls and women of this era often gathered in small sewing groups, which fostered social cohesion and close friendships in the process.

### Quilts

At their most basic, quilts are simple bedclothes made of batting sandwiched between pieced-together homespun. However modest in construction and appearance, quilts were absolutely necessary in most parts of the country for keeping people warm. At quilting bees, women would gather to turn necessity into wondrous inventions. Quilters would piece cloth together into artistic designs and perform fine needlework to fix the various components of the quilt in place. The top of the quilt held the design, which itself often comprised separate smaller designs, or blocks, each brought to the bee by a participant. Housewives tended to have clothing and furnishings of the same color, since fabric had to be purchased by the bolt. At bees, they would trade and combine their fabrics to make multicolored designs—"crazy quilts." Friends and neighbors contributed whatever scraps of fabric they had saved for such a purpose. Nothing was wasted—snippets left over from the making of a garment went into the "piece bag." So precious was cloth that piece bags would be passed down from mother to daughter. Women memorialized their collective experiences and individual stories through their quilts, much as scrapbookers and other homemaker-artists (including the legions of contemporary quilters) do today. As a woman rhapsodized in 1845,

> Yes, there is the PATCHWORK QUILT! looking to the uninterested observer like a miscellaneous collection of odd bits and ends of calico, but to me it is a precious reliquary of past treasures; a storehouse of valuables . . . a bound volume of hieroglyphics, each of which is a key to some painful or pleasant remembrance . . . it contains a piece of each of my childhood's calico gowns, and of my mother's and sisters'; and that is not all.[186]

The romanticized version of big quilting bees, around which exuberant, community-wide parties formed, was not the norm. But the gatherings regularly brought together groups of four to eight women (four to a quilting frame). They would share light, the warmth of a fire, perhaps a meal or a baked treat, and the stimulation of talking to one another, all the while accomplishing a vital chore:

> One day a week, when the neighbors came to quilt, my brother would take the bed in Mama's room down to the

kitchen and put up the frame for that day. It was quite a job, but he never minded . . . Dad was always proud of Mama on quilting days. When he came inside from work he would say how busy she had been. He knew that she had a hard and lonely life . . . he was glad she could have a day with her friends and enjoy herself.[187]

Then, as now, loving husbands recognized that a woman's time with her friends added to her sense of well-being and thus improved family life.

Sometimes groups of quilters did throw a big bee, to which men were invited, usually in the evening. Often, these parties centered on finishing a young woman's last wedding quilt— the thirteenth after she alone had made the first twelve. Quilt by quilt, the designs and techniques of the first dozen gained in complexity. When the girl had reached the thirteenth, the "Bride's Quilt," she and her friends would throw a special bee and join in the quilting. One archival quilt memorializes a warning that girls probably heard all too often:

*At your quilting, maids, don't dally,*
*Quilt quick if you would marry,*
*A maid who is quiltless at twenty-one*
*Never shall greet her bridal sun!*[188]

With advances in the textile industry by the 1840s, quilting often evolved into a leisure pastime rather than a necessity. Thus the friendship-quilt fad came to life and swept the nation. The hallmark of friendship quilts was signatures. A quilt maker would ask each of her girlfriends to contribute a block as part of the overall design of her friendship quilt. Included in each block would be a piece of fabric that bore the friend's signature, in either indelible ink or embroidery. Important dates might be included with the friend's name, often the words "remember me," and sometimes a line of verse.

Many women about to be torn from their families and friends when their husbands and fathers decided to join the great westward migration received friendship quilts as going-away gifts. In faraway places, friendship quilts were displayed as comforting reminders of important connections, and often they were part of the matriarchal legacies passed down through generations.

Quilts by skilled makers often achieved artistic excellence, whether created as objects for display or as useful providers of warmth. In the latter category, wonderful examples were created by African American women who learned the practices of quilt making when they were slaves and passed these skills down through the generations. As part of the same cultural transmission, these quilt makers also came to cherish the cohesion and friendship that emerged from quilting together. In a rural hamlet south of Selma, Alabama, the Gee's Bend Quilters originally started making quilts from feed sacks and cast-off work clothes. Their skilled hands yielded cozy coverlets to get their families through the night. But as they worked together, chatting and sometimes singing, these quilters evolved a distinctive style inspired by West African textiles and modern geometric paintings. Today, these communal objects are now regarded as high art. The Gee's Bend quilts have been featured in museum exhibitions, on television, and on 2006 US postage stamps. Today, collectors vie at auction for such one-of-a-kind masterpieces.[189]

## Church Groups

In the early 1800s, new groups of women were formed by the rampant Christian fervor that was subsequently dubbed the Second Great Awakening. Throughout the sparsely populated Midwest, religious camp meetings offered both worship and fellowship to enthusiastic men, women, and children of all ages. These revivals afforded women opportunities to learn public speaking and organizational skills, and display their culinary mastery during Sabbath potlucks.

At mid-century, Harriet Walter, a Kansas pioneer, remembered how Mary Clarke's kitchen bustled with communal activity in advance of a Sabbath meeting:

In Ernest Clarke's house all was excitement, for Brother Craft, the Baptist Preacher, would be at the School house Saturday afternoon for the monthly covenant meeting. Mary Clarke carefully set her sponge for the white bread Friday evening . . . the Boston brown bread was steaming too. Of course, there were cakes to be baked and beans and rice pudding if there were no pies . . . there must be meat too. In summer the ham . . . or the chicken . . . in

winter the large beef or pork roast was cooked to a turn
. . . she scrubbed the kitchen to shining cleanliness.[190]

Food, traditionally "woman's sphere," made them indispensible to the revival movement, while offering them opportunities to bond outside their homes.

At prayer meetings, people were called upon to embrace the revivalist approach to eternal salvation by bearing witness, with great emotion, in front of the crowds. Women were welcome to express their conversions along with the men. Thus, under the auspices of religious piety women took the floor in public meetings.

As regular adjuncts to the revivalist meetings, female-only prayer groups began to proliferate in more settled areas. For such an unassailably righteous motivation, women were allowed to engage with one another outside the sphere of their families. These prayer meetings rapidly morphed into ladies' "cent" societies, to which members contributed whatever "mite" they could spare to raise funds for missionary efforts and the distribution of Bibles. By the end of the century women were participating in "mission boards," which raised funds to prepare women missionaries for overseas work. In 1915, at the height of the missionary fervor, an estimated three million women were members of foreign mission societies.[191]

Throughout the nineteenth century, the Industrial Revolution reshaped daily activities, thereby freeing middle-class women from some of the raw-knuckled labor that had been required of their forebears. Churchgoing and associated activities gave women an out from home confinement and allowed them to build friendships in the service of a moral cause.[192][193]

Many of these church-related groups soon repurposed themselves as engines of social change. Observing the acute dislocations caused by rapid growth in their communities, the women of church auxiliaries did not limit themselves for long to distant missionary causes or Bible distribution. Typical women's church societies ministered to orphans, the poor, and unwed mothers. For groups of middle-class women, sewing fund-raisers replaced the quilting bees that had been based in immediate family needs.

In 1839, twenty-nine women who attended the Calvinist Church in Worcester, Massachusetts, formed the Centre Missionary Sewing Circle. They were determined to raise money for foreign missions in order to "win over unhappy subjects of the Prince of darkness." The women met twice a month from two to nine p.m., rotating among members' homes. They sewed both plain and "fancy" work for sale. The Circle's detailed records indicate that they acquired considerable merchandising and marketing expertise, which over the course of a few years boosted sales a hundredfold. The group grew rapidly, along with the town (to seventy members in three years), and as new members joined, they became friends and colleagues.[194]

Church-related women's groups matured into established institutions—every community had at least one, often several. These associations gradually gained in confidence and scope. Soon women were co-opting men's "rules of order" to conduct their meetings. Many churchmen grew alarmed that a group of women was making organizational decisions without male supervision and attempted to control their activities.

In Worcester, when the Sewing Circle decided to redirect its efforts from supporting distant missions to focusing on serving the poor of their own town, the Calvinist minister declared himself "decidedly against" the plan. After months of dithering about whether to submit to the minister, the women voted to strike out in the direction they had chosen for themselves. This vote constituted a momentous act for the middle-class members, who no doubt regarded themselves as paragons of Victorian domesticity. They did not stop with this one act, however. In what amounted to a formal nose-thumbing at the minister's meddling, the women also voted to amend their constitution to state their goal of aiding the local poor: "How much more good might we accomplish by engaging heartily in the work before us than by wishing for other fields of labor?"[195]

Despite the kerfuffle in Worcester, it was not a craving for power that drew great numbers of women to their church groups. Ostensibly, their gatherings were tied to some religious mandate. But what really got them out of the house, putting their best foot forward, and eager to attend was the social balm the groups provided.

The revivalist movement of the early nineteenth century held out a message of personal agency in one's salvation. Women extended this self-empowerment to a social domain that allowed them to meet one another outside domestic spheres. In turn, the daughters of this generation of revivalists marched together in the great wave of societal reform that swept through the country at midcentury. Female friendships were thus freed from the confines of home and thrust into the larger world.

### Early Reform Groups

One of the most powerful of the early reform groups was the New York Female Moral Reform Society, formed in 1834 and dedicated to the Sisyphean task of eliminating prostitution. Prostitution was not wholly illegal at the time, and up to 10 percent of women in New York generated income from the practice. Within five years, the New York group had mushroomed into 445 auxiliaries and changed its name to the American Female Moral Reform Society. This formidable organization butted heads with the male establishment by lobbying legislators to make sex solicitation by men illegal and, perhaps more compellingly, by threatening to publish the names of brothel frequenters. Whatever the efficacy of its efforts, the energy and organizational muscle of the NYFMRS laid the groundwork for the great surge in women's progressive movements later in the century.

In Philadelphia and other urban centers, women recognized that working-class and poor women were the most vulnerable to economic hard times. This fact was not lost upon the women who actually lived in the underbelly of urban areas, or near its edge. These laboring women, including many free African Americans, formed mutual aid societies to help their members survive.

The Daughters of Africa, about two hundred strong in the 1820s, chronicled their activities in an Order Book, which recorded not only the Daughters' charitable activities but also their organizational emphasis on respectability.[196] With the odds for a decent life stacked so heavily against black women of that time, it certainly was easier to cultivate refinement when joined in the effort by one's friends and peers.[197]

### Frontier Friendships

Women isolated on the far western fringe of the US territories were in dire need of friends. Their clubs in the Pacific Northwest developed later than did women's clubs east of the Mississippi, simply because there were so few women "out there" in the early days of frontier settlement. As soon as real villages with families at their cores took root, western women's organizations spread like one flame to the next in a candle-lighting ceremony, starting with prayer groups and segueing into social reform.

Settlements in the West faced a yawning void of civic leadership. Well-educated ladies were among the pioneers, and they, along with everybody else, confronted streets without sidewalks, steaming piles of garbage infested with flies and rats, tainted water, no sanitation, no schools, and certainly no libraries. Many of these ladies possessed organizational and leadership skills gained from club activities in their eastern hometowns, and they pooled their talents and resources to catapult their new organizations into civic and political leadership. Out west, this happened in a compressed time frame compared to that in the East.

In 1838, a group of women in the Pacific Northwest, isolated in a dangerous, wild land, came together as the Columbia Maternal Association, which was among the very earliest women's organizations on the frontier. These women were the wives of missionaries. At the outset, only two of the original six were actually mothers, but they had a bevy of orphans on their hands due to the appalling death toll of the Oregon Trail. These women, whose number soon grew to twelve, could meet only rarely, but using the means at hand they circulated reading materials for discussion. Reflecting the compressed time frame of club development, their charter moved from a religious focus in one clause to a highly practical mandate in the next: "[Each member will] qualify herself by prayer, by readings & by all appropriate means, for performing the arduous duties of a Christian mother, & suggest to her sister members such hints as her own experience may furnish or circumstances render necessary."[198] Imagine what these hints might have included—birth control strategies, laundering tips, opiate and medicinal recipes, proper behavior when confronted by a scantily dressed Native American man in one's kitchen. Significantly, each member of the Columbia Maternal Association also pledged to look after one another's children should the mother die. This pact exemplifies female friendship at its most fundamental level.

To pioneer women, even those who had once lived placid, middle-class lives, the Victorian ideal of the angel in the house must have seemed as unattainable as supermodel stardom does to the average American woman today. Real life did not give scope to such daydreams. A wonderful, if lurid, episode that occurred in 1846 tells of two women who saved each other and their families during a desert crossing in the territory that is now New Mexico. Two families—the Benhams, husband, wife, and seven-year-old; and the Braxtons, a couple and two teenage sons—were heading for a homestead near the Pecos River. Their last water cask burst as they were crossing the Staked Plain, and a rattlesnake bit one of the boys. The men developed sunstroke and things looked as bad as could be—until Murphy's Law prevailed and a heavily armed band of Mexican marauders attacked them. The bad guys robbed the families of what little they still carried, killed their horses and mules, and left the men and children to die. Unfortunately for the bandits, they abducted the women.

Locked in an upstairs room in the outlaws' hideout, Mrs. Benham and Mrs. Braxton yanked off the wooden window bars, dropped to the ground, and stole all the fresh horses, after cutting the hamstrings of the horses that were resting. They were not messing around! They succeeded in rescuing their families, who went on to homestead in New Mexico.[199] We may interpret what happened to these two as women's friendship writ large—a battlefield bond that would last a lifetime. Would either have made it so far without the other?

### Transcendentalist Friendships

In the early nineteenth century, many women with access to books immersed themselves in self-education. Public education, feeble though it was, gave girls even shorter shrift than boys. For the most part, private secondary schools were male-only. Girls with a thirst for knowledge began to form women-only groups for intellectual discussion. Often, these groups focused their efforts on the Bible or other religious texts. Then, spurred by a handful of remarkable outside-the-box female leaders—including Elizabeth Peabody and Margaret Fuller—women began to meet for the express purpose of expanding their knowledge.

Elizabeth Peabody (1804–94), one of three remarkable sisters, gained fame for her liberal views on education. She found herself near the center of an intellectually intense circle of earnest young transcendentalists with names we readily recognize today, such as Ralph Waldo Emerson, Horace Mann, Nathaniel Hawthorne, and Margaret Fuller. So that this group could have ready access to the latest writings of European romantics and transcendentalists, Elizabeth established a bookshop in Beacon Hill, Boston, where the intelligentsia of "the Athens of America"[200] would gather. In 1832 she launched a series of reading parties for women, held at her bookshop. These gatherings involved readings, lectures—many given by the vivacious Elizabeth Peabody herself—and discussions of great works of the Western canon, ranging from the ancient Greeks to the French Revolution.

Elizabeth's friendship with her brilliant younger contemporary Margaret Fuller (1810–50) was fraught. Peabody, in the manner of a mentor, showed many kindnesses to Fuller. Not least, in 1838 she made her bookshop space available to Fuller so that Fuller could conduct her own adult education classes, "Conversations," for women. These offerings, for which participants paid a small fee, built upon the foundation laid by Peabody, who was highly complimentary of Margaret's efforts: "Miss Fuller's thoughts were much illustrated, and all was said with the most captivating address and grace, and with beautiful modesty."[201]

On the other hand, Fuller, who had by then eclipsed Peabody in her public stature, did not treat her benefactress with similar respect. In fact, Fuller made fun of Peabody behind her back. William Ellery Channing, a leading Unitarian theologian, chided Fuller for her unkindness: "When I consider that you are all that Miss P[eabody] wished to be, and that you despise her, and that she loves and honors you, I think her place in Heaven must be very high."[202] Margaret Fuller, the "queen bee" of her day, could sting as sharply as any twenty-first-century "mean girl."

Even so, she championed women generally in her groundbreaking manifesto, *Woman in the Nineteenth Century* (1844). There she took a stance that defied the patriarchal mores of her day: "I would have Woman lay aside all thought, such as she habitually cherishes, of being taught and led by men."[203] Fuller's public Conversations encouraged women to exercise their intellects. As she wrote to her friend Sophia Dana Ripley on the nature of the proposed Conversations:

The advantages of a weekly meeting, for conversation, might be great enough to repay the trouble of attendance, if they consisted only in supplying a point of union to well-educated and thinking women, in a city which, with great pretensions to mental refinement, boasts, at present, nothing of the kind.[204]

Fuller embraced the idea of female camaraderie going hand-in-hand with intellectual exploration, an idea that soared with subsequent generations of college women.

She did try to open up her Conversations to men, but the result was a quashing of the women's spirit: "The men took over the discussion and performed for each other," she noted. She immediately reverted to the earlier, successful, format, in which the Conversations reflected "a decidedly feminine character, feminine concerns addressed in feminine language in an atmosphere of feminine intimacy."[205]

Yet both Fuller and Peabody nurtured platonic relationships with the leading male transcendentalists of their day, friendships based in mutual esteem. Fuller proposed a revolutionary vision of human sexuality that contradicted the prevailing Victorian doctrine of separate spheres for men and women: "Male and female represent the two sides of the great radical dualism. But, in fact, they are perpetually passing into one another. Fluid hardens to solid, solid rushes to fluid. There is no wholly masculine man, no purely feminine woman."[206] In this androgynous perspective, Fuller was way ahead of her time. Her view would ring true to many today.

Of course, where boundaries are blurred there is room for ambiguity. Ralph Waldo Emerson and Fuller wrote to each other about the nature of friendship, with Fuller often mixing the concepts of love and friendship. Both were vexed, however, by the difficulties they had being comfortable with each other face-to-face. Fuller complained to her confidante, the romantic poet Caroline Sturgis, of the "perpetual wall" that kept her from getting close to Emerson in person, and sometimes even in their letters.[207] When Margaret wrote to Waldo that "I am yours & yours shall be," her intensity caused him to slam on the brakes. He explained: "You & I are not inhabitants of one thought of the Divine Mind, but of two thoughts . . . essentially unlike."[208] The complexities of their relationship never allowed them to achieve emotional reciprocity.

All Margaret Fuller's relationships can be described as "complicated." Her commitment to androgyny permeated her friendships and her loves. When she was twenty-one, she met the somewhat younger, captivating Anna Barker and embarked upon what was then called a romantic friendship. Fuller wrote of her new obsession: "I loved Anna for a time with as much passion as I was then strong enough to feel—Her face was always gleaming before me, her voice was echoing in my ear, all poetic thoughts clustered round the dear image."[209]

When she was twenty-eight, Fuller fell in love with Samuel Ward, eight years her junior. After a dalliance, which apparently meant less to him than it did to her, he drew away, and she played the part of a rejected lover: "If you love me as I deserve to be loved, you cannot dispense with seeing me." Whatever sort of "love"—friendly, worshipful, or erotic—she had in mind, she could not have been pleased when he told her he thought of her more as a mother figure. Shortly thereafter, she learned he was in love with none other than her earlier love, Anna Barker, whom he married.[210]

Margaret's dearest friend throughout her turbulent career was Caroline Sturgis, another figure of stature among the transcendentalists. The two women traveled and visited mutual friends together and, at various intervals, lived together, including an idyllic summer spent in a beach house. Fuller's letters to Caroline intertwine gossipy observations with lofty intellectual musings. Attesting to the matching of their minds as well as the knitting of their souls, in 1844 Margaret reminded Caroline of a shared memory, both cuddly and soul moving, which she compares to one of the few female friendships recorded in the Bible: "Do you remember that night last summer when we fell asleep on the bed and we were like Elizabeth and Mary. I have often wanted to express what appeared to me that night, but could not, only every day I understand it better. I feel profou[nd]ly bound with you and hope you wear my ring."[211]

These three examples show that Margaret Fuller not only "talked the talk" but also "walked the walk" of androgyny. In her love for Anna Barker, Samuel Ward, and Caroline Sturgis, she followed her passions regardless of gender.

It was not out of character that she pulled up stakes at age thirty-six to voyage to Europe as a foreign correspondent for Horace Greely's *New-York Daily Tribune*. In Italy she covered the revolutions of 1848 to 1849. She took as a lover an Italian officer of the Roman Guard and with him conceived a son. The small family sailed back to the United States in 1850, only to die in a violent storm off Fire Island. Henry David Thoreau and other devastated transcendentalists combed the shore for days, seeking her body, but to no avail.

### Inside the Lowell System

At the same time that the intellectual elite clustered within the American transcendental movement, less privileged women found themselves swept into new relationships by the inexorable tide of the Industrial Revolution. Fresh-faced swarms of immigrants and farm girls, generally age fourteen to twenty-five, joined the working classes in urban areas. For the most part these youngsters came from working-class families, many of which needed the girls to support themselves and contribute to strained family resources.

Lucy Larcom (1824–93) began work in a cotton mill in Lowell, Massachusetts, when she was eleven years old. Although child labor is no longer acceptable today, the Lowell system was progressive for its day in that it provided mill girls with supervised dormitories, a grammar school, numerous night schools, cultural events, Bible study groups, and libraries.[212] Lucy Larcom, who later in her life became a prominent poet and writer, described in her memoir, *A New England Girlhood*, what it was like for girls to spend their formative adolescent years at the heart of a radical social experiment. From her writing and those of other mill girls, we know that they made the most of the friendship opportunities that presented themselves. Lucy wrote of her girlfriends:

I regard it as one of the privileges of my youth that I was permitted to grow up among those active, interesting girls, whose lives were not mere echoes of other lives, but had principle and purpose distinctly their own . . . They were earnest and capable; ready to undertake anything that was worth doing. My dreamy, indolent nature was shamed into activity among them. They gave me a larger, firmer ideal of womanhood.[213]

Such give-and-take could describe the value of girlfriendship in any age.

### Late Nineteenth-Century Working Women

Near the end of the nineteenth century, girls flocked to cities for a chance to raise their socioeconomic status. The paternalistic model of the Lowell system was long gone, a victim of competitive economics. Strict working conditions on factory floors and in stores impeded conversation among workers. And most girls did not live in quarters that could easily accommodate visitors after work. Factory girls formed their own clubs, sometimes with the sponsorship of upper-class, reform-minded women. From the sponsors' points of view, these clubs served to guard girls' virtue by offering alternate gathering places to saloons and door stoops. Moreover, the urban working women's clubs provided what the girls really needed: emergency loans and medical benefits; an employment agency; bathrooms with warm water; cheap, hot meals; and job-advancement classes in typewriting, spelling, and other office skills.

The National League of Women Workers, established to coordinate the activities of all working women's clubs, published a monthly journal, *Far and Near*, which cited sociability as the major attraction of club membership. One member of the Good Will Club of Brooklyn noted that the girls who forged the closest "heart relations" with other members were those who led the others in self-improvement. True friends could dissuade their club sisters, for the sake of the group's respectability, to avoid vulgar behavior such as loud talking, slang, gum chewing, ostentatious clothing, and bold flirting. The girls of the urban working clubs took their efforts to better their lots extremely seriously, and a great many of them did indeed climb the social ladder, with the support of their club friends.

The 1894 convention of the Association of Working Girls' Societies issued this plea, emphasizing the importance of friendship among its members: "Let us encourage social life, for it is the centre, the heart, of the club; upon it the very existence of the club depends, for unless a club is social it is not co-operative, and unless it is thoroughly co-operative it is not successful."[214] Irene Tracy, one of the members from the Thirty-Eighth Street Working Girls' Society, expressed the psychological benefits she derived from spending the evening at her club:

> I was feeling rather low in my mind last night and thought
> I would go to the Club, which I did. I can hardly explain
> just what it is that makes me feel better . . . it seems the

pleasant smile or jolly word or two . . . but above all, I think, it is the oneness . . . some of our members don't have to go out and earn their daily bread; we all meet on the common ground of womanhood and sisterhood; we mutually bridge over a chasm . . . the Club working girl does not feel that she is looked down on, but feels she has gained the respect, love, sympathy, and loyalty of a stanch [*sic*] friend, while the woman of leisure feels she has gained a true friend in the girl who has to go out into the world alone, who has learned so well how to help herself, and is such a true, womanly woman; for there is something strong and self-reliant about her; she is to be trusted.[215]

### The General Federation of Women's Clubs

At the same time that urban working women were joining clubs to improve their social status, middle-class women also started joining clubs in droves. By 1900, there were thousands of women's organizations studded across the nation. All these groups shared the goal of self-improvement. Why not improve one's mind as one sat in the amiable company of women like oneself? These meetings were not to be confused with sewing circles or quilting bees. In fact, to encourage undivided attention, some clubs actually forbade knitting or sewing during lectures and discussion. Most clubs were small enough to meet in members' parlors, but in large cities, where membership in a single club could reach into the hundreds, they often rented public rooms, and a few, like

San Francisco's elite Town and Country Club, managed to buy their own clubhouses.

Jane Cunningham Croly, among the pioneers of the women's club movement, sought to bring shared wisdom to clubs across the country by founding the General Federation of Women's Clubs in 1890. Local branches of the GFWC appealed particularly to high school and college graduates, who would meet to discuss literature, the arts, and issues of social reform, such as the establishment of kindergartens, care for the elderly, and public health facilities. In clubs like the Women's Municipal Leagues of Boston and New York, women with similar concerns found one another and worked together on progressive causes, which often led to long-term alliances and intimate friendships.

The history of the Chicago Woman's Club is illustrative. After an initial period from 1876 to 1883 when members gathered together to discuss books and "suitable" topics (women's suffrage was not one of them!), the group debated on December 5, 1883: "Shall Our Club Do Practical Work?" The vote was affirmative, and members banded together to provide the money, materials, and teacher for a pioneer kindergarten. Then, in the late 1890s, the club took the lead in forming what was known as the Vacation Schools Committee, which garnered the assistance of sixty different women's organizations to give city children a summertime experience in the country.

Yet however progressive, these women did not extend club membership or friendship possibilities to nonwhite women. African American women, rebuffed in their efforts to join established women's organizations, created their own societies. Many black churches had women's literary societies, such as clubs in Brooklyn and Philadelphia, where it was common for women to write poems and essays that were circulated among members, fervently discussed, and politely criticized. In 1892, black teachers from Washington, DC, founded the Colored Woman's League, and in 1904 the National Association of Colored Women's Clubs was incorporated to support black women's "moral" and "material" progress. Like many white women's clubs, these organizations campaigned for temperance and suffrage, but they also worked specifically to improve education for people of color, assist Southern blacks to migrate from the South to northern cities, counteract Jim Crow laws, and provide a meeting space where black women with similar concerns could find one another and become friends.

The New Era Club, started by Josephine St. Pierre Ruffin in Boston, was a prominent organization for progressive black women. With the assistance of her daughter, a graduate of Boston Teachers College, Josephine edited the *Woman's Era*, a monthly magazine. Club members were urged to submit material on such subjects as literature, music, suffrage, antilynching, temperance, and prison reform, in addition to news of births, graduations, marriages, travels, and cultural events.[216]

Ethnic organizations, such as the Polish Women's Alliance and the Italian Women's Civic Club, offered a welcoming environment to the hundreds of thousands of immigrants who came to America from Europe around 1900. The Polish Women's Alliance, founded in Chicago in 1898, was particularly successful in helping both working-class and educated women join forces to assimilate and succeed. The editor of their newsletter put it this way: "Let us join hands—women who do hard labor and women of words and thoughts—let us believe in each other . . . [let us] create what Polish women want and desire."[217] The organization had grown to 1,400 members by 1903.

In New York City, the Young Women's Hebrew Association, founded in 1902 on the model of the YMCA and the YWCA, provided residential accommodations to more than one hundred working-class women, and also offered gym facilities and swimming classes for others living in the neighborhood. These organizations gave women a sense of belonging to a familiar group, where members could lapse into their mother tongues even as they learned English and struggled to become American.

### Susan B. Anthony and Elizabeth Cady Stanton

In the pantheon of women's friendship, the legendary bond between Susan B. Anthony (1820–1906) and Elizabeth Cady (1815–1902) Stanton stands in a class by itself. In 1869, Anthony and Stanton founded the National Woman Suffrage Association. Membership was restricted to women, in the belief that if men were accepted they would come to dominate the association. The two friends were so committed to women's right to vote that, although they had both staunchly supported abolition, they refused to support the Fourteenth and Fifteenth amendments, which granted suffrage to African American men long before American women got the vote. In 1890, the NWSA and another major suffrage association (with members of both genders) became the National American Woman Suffrage Association, which continued to promote the vote for women until it was finally passed with the Nineteenth Amendment in 1920. By then both Anthony and Stanton were dead, Stanton in 1902 and Anthony in 1906. But the legacy of their personal and political friendship accomplished its goal.

Anthony and Stanton first met shortly after the latter had promulgated "The Woman's Declaration of Independence" at the Seneca Falls and Rochester, NY, conventions of 1848. At that time, Stanton was already the mother of four children— she would eventually have seven—and Anthony was thirtyish and single. Their mutual interests in antislavery, temperance, and women's rights provided the bedrock for a friendship that would last for more than half a century. In her autobiography, *Eighty Years and More*, Stanton told her readers that since Anthony was so closely connected to her personal narrative, it would be necessary for her to tell Anthony's story as well as her own. Two chapters and countless mentions of Susan B. Anthony spell out their devotion to each other and to the causes they championed together.

Stanton remembered: "Miss Anthony and I wrote addresses for temperance, anti-slavery, educational, and women's rights conventions . . . we forged resolutions, protests, appeals, petitions, agricultural reports, and constitutional arguments . . . we made it a matter of conscience to accept every invitation to speak on every question, in order to maintain a woman's right to do so."

Despite their different situations in life, different temperaments, and different strengths, Stanton asserted that she and Anthony never experienced the slightest falling-out. "So entirely one are we that, in all our associations, ever side by side on the same platform, not one feeling of envy or jealousy has ever shadowed our lives. We have indulged freely in criticism of each other when alone, and hotly contended whenever we have differed, but in our friendship of years there has never been the break of one hour."

In their private relations as well as in their public appearances, there seems to have been remarkable harmony between the mother of seven and her single friend. Stanton referred to Anthony as her children's "second mother" and her own "good angel," as she fondly recalled the times when her friend would take one or two of the Stanton children to the Anthony family farm outside Rochester.

Over the years, Anthony and Stanton traveled together, not only to conventions and state legislatures throughout America but also to Europe, for both professional reasons and cultural enjoyment. With scarcely a nod to her husband, Stanton had no qualms writing that she felt wedded to Anthony, almost like they were husband and wife: "So closely interwoven have been our lives, our purposes, and experiences that, separated, we have a feeling of incompleteness—united, such strength of self-assertion that no ordinary obstacles, differences or dangers ever appear to us insurmountable."[218]

Theirs was a remarkable story of friendship between two brave women who were definitely ahead of their times. Had they been born during the last quarter of the nineteenth century, instead of the first quarter (Stanton was born in 1815, Anthony in 1820), they would undoubtedly have become "New Women," with college educations and paid employment. Had they been born in the mid–twentieth century, they would have become second-wave feminists supporting such causes as women's reproductive rights and the Equal Rights Amendment. As it was, they found in each other the perfect "friend and coadjutor" to combat the prejudices of their age and fight for women's rights. No longer confined to the "soft whispers of friendship," as had been Mercy Otis Warren and Abigail Adams, they stood side by side in the public glare— one skinny and angular, the other plump and curly-haired. And their friendship, intimate and unique as it was, served to ignite group friendships around issues of vital importance. Stanton and Anthony lofted Aristotle's ideal of friendship as a civic virtue to one that included women on the national stage.

*⁀*

NINE

# COLLEGE GIRLS, CITY GIRLS, AND THE NEW WOMAN

*"The Hull-House Woman's Club was one of Jane Addams's favorite endeavors. It brought together women from all over the world. Once a week the women could leave their dreary homes and commune with other women and enjoy the hospitality of a cup of tea and a piece of cake."*

—HILDA SATT POLACHECK, *I CAME A STRANGER: THE STORY OF A HULL-HOUSE GIRL* (EARLY TWENTIETH CENTURY)

*"Neither of us had ever known any pleasure quite equal to the joy of coming home at the end of the day after a series of separate varied experiences, and each recounting those incidents to the other over late biscuits and tea."*

—VERA BRITTAIN, *TESTAMENT OF FRIENDSHIP*, 1940

*G*ODEY'S LADY'S BOOK—THE FOREMOST AMERICAN journal for women during much of the nineteenth century— ceased publishing in 1878, and attempts to resuscitate it by steadfast supporters were unsuccessful. Its prescriptions for feminine purity, self-sacrifice, modesty, and dependence were no longer attractive to the "New Woman," who had been shaped by higher education, employment, city living, and relative freedom. Although most women still looked to marriage and motherhood for their chief occupations in life, a new breed of restless women was bent on enlarging the scope of their pursuits, and consequently changed their friendship patterns. Whereas girls and women had once looked to their sisters, cousins, and immediate neighbors for their closest friends, the New Woman would find hers among schoolmates and college chums, in women's clubs, at the workplace, and in urban environments often far from their hometowns.

The New Woman was both a European and an American phenomenon. She was to be found in London, Paris, Berlin, Stockholm, and Moscow as well as New York, Boston, Chicago, and San Francisco. She was characteristically young, well educated, high spirited, dynamic, competent, and daring. As pictured in the many images drawn by Charles Dana Gibson for *Life* magazine during the 1890s, she was likely to be attired in a high-collared white shirtwaist tucked into a dark, comfortable skirt that stopped at the ankles. Advertisements, magazines, and posters promoted the image of the New Woman, just as other forms of mass media would later exhibit images of the flapper, the housewife, the wartime worker, and the androgynous feminist. The bicycle was the symbol of the New Woman's freedom outside the home, as she raced off with her friends—men or women—down city streets and into the countryside. However limited in actual numbers, the image of the New Woman set the tone for and affected the lives of countless American women.

There is no doubt that increased opportunities for education contributed to the formation of the New Woman. She and her college friends were among the first and second generations of American women to reap the benefits of changes that had taken place during the nineteenth century, when colleges and universities, originally restricted to male students, began to admit women—Oberlin as early as 1837. While traditional colleges in the East remained exclusively for men, many new colleges and universities in the Midwest and West began as coeducational institutions or eventually became coed. At the same time, the creation of women's colleges, such as Vassar in 1861, Wellesley in 1870, Smith in 1871, and Bryn Mawr in 1885, added a new level of rigor and prestige to women's higher education—for those who could afford it. By 1900, there were eighty-five thousand women undergraduates.[219]

Once graduate schools opened to women, enrollment rose exponentially. In 1890, 10 percent of graduate students were women; by 1918, that figure had risen to a startling 41 percent.[220] Some of the most studious undergraduates were encouraged to take graduate degrees in the humanities and sciences and then to seek academic positions in women's colleges. A very few were funneled into medicine and law, fields that were newly, if grudgingly, open to women. Whereas undergraduates usually had a large number of "college girls" in their pool of potential friends, those who went on to graduate or professional schools often found themselves with only one or two other women in their disciplines—all the more reason for those rare birds in the uppermost echelons of higher education to perch together and remain close friends.

## College Friends

Linda W. Rosenzweig, who has studied friendship in the lives of American women beginning with college students around 1900, notes the persistence of intense affection between them but also a worrisome tone in "the ideology of friendship" as expressed by journalists and policy makers. Indeed, a spate of articles between 1900 and 1920 lamented the death of the "art of friendship" among men derived from the fear that personal relationships were, in the urbanized, anonymous twentieth century, becoming little more than mere acquaintanceships. Yet other articles that centered on women's relationships stressed the benefits of friendship as an educational, ennobling, and personally satisfying experience. One stated emphatically: "We live for our friends, and at bottom for no other reason." Another asserted that "in the emotional region, many women, but very few men, can form the highest kind of tie"—that is, true friendship. By the time this last statement was written, the concept of friendship easily accommodated the female gender; it had been adjusted to the emotional largesse associated with women rather than the restraint and reason expected of men. Yet not everyone agreed with this positive view of women as friends. Some continued to declare that "the sisterhood of women is inconceivable" and that women were "naturally treacherous to one another." Old prejudices die hard.

Within this fraught atmosphere, professors cautioned women students to make friends carefully, to seek out only those who would lift them up intellectually and morally, and to avoid those who would bring them down or lead them astray. A 1901 member of the Smith College faculty counseled college students to form friendships inspired by lofty ideals and high-culture art, with the aim of loyalty to one another. Comfort was to be found inside the college "crowd," with its sense of community and sisterhood, rather than in a frenzied search for the elusive man or, perish the thought, "a premature love affair," as another counselor put it.

The sheer number of colleges and universities open to women promoted friendships between female students, primarily from the middle and upper classes. Their letters and diaries suggest that friendship was what many of them valued most in their college experiences. Dorothy Mendenhall remembered from her years at Smith that "more than actual study, my contact with girls my own age and the making of a few close friends, were the most valuable part for me." Later, after she had graduated and started medical school at Johns Hopkins in Baltimore, she linked up with a former classmate whom she hadn't known well at Smith but who "became a dear friend" because they had similar college experiences and professional interests. One can imagine how much these two women must have needed each other as they faced the overbearing male medical establishment, which permitted only 5 percent of incoming students to be female.

Jane Cary at Wellesley also appreciated the classmates who made her life "fuller, richer, and happier." She remembered all-female campus parties, where younger students routinely asked upperclasswomen to dance with them; weeks could go by without her ever seeing a man. In the letters she wrote to her mother from the years 1913 to 1914, she commented regularly on her relationships with other Wellesleyites. Sometimes a new friendship developed unexpectedly, as in the case of Helen, who entered her life unbidden: "I never would have thought of her for a friend if she hadn't been so nice to me and done all the making friends part."

College girls found time for friendship at meals, at the gymnasium, while walking to and from class, during concerts and plays, and while holing up in a dorm room for late-night chats. They discussed the personalities of their professors, the books they were reading, the sermons they heard at chapel, the latest fashions, and topics from the news, such as women's suffrage. They confided in one another, as women always had, their interest in particular men as well as the details of their friendships with other girls, which were sometimes riven by rivalry and jealousy. Most of the women would end up marrying, and many would share with one another their wedding plans, down to their choice of rings, wedding gowns, flowers, and honeymoons. Many would continue to be friends after college as they became wives, mothers, and, later in life, widows.

Annie Sears, a kindergarten teacher, never went to a

formal college, but she nevertheless became friends with her sister's college friend, Frances Rousmaniere, who pursued doctoral work. From Waltham, Massachusetts, Annie wrote to Frances in Cambridge, expressing her ardent wish for them to remain friends despite the geographical and educational distance between them. She reflected on the importance of keeping in touch with friends in her letter of October 1, 1903: "While I agree with you and Emerson that often we must live on trust in regard to our friends, yet experience seems to show me that if friends never meet or exchange a line for a long time, so they do inevitably in spite of themselves drift apart." Though Frances went on to teach at Mount Holyoke College and then to marry, her friendship with Annie survived.

Annie's letters indicate that she was able to share, vicariously and sequentially, Frances's life as an academic, new bride, and mother. She wrote after her first visit to Mount Holyoke: "I am glad that I can picture your life so much better now, both at your classes, in your apartment and at your jolly little meals." At the news of Frances's engagement, she exclaimed: "You cannot think how much I wish I could see you, and know a little what this new life means to so dear a friend." But she also wondered if it would be hard for Frances to give up her job as a professor and her "semi-Bohemian college life": "You know that I am thinking of you all these days . . . wishing all good wishes, praying that it may be truly the best life for you."

Annie's reflections on the exciting times they lived in, when women had so many more options than in the past, was also tempered by a realistic sense that "the world is not made for women": "There seems to be unhappiness in store whether we belong to the old type [of women] or the new type." As independent teachers, Annie and Frances definitely belonged to the new type. We don't know if either one bobbed her hair or smoked cigarettes in public—shocking hallmarks of the New Woman in the early twentieth century—but we do know that they continued to prize their long-term relationship. In 1914, when World War I was about to destroy the lives of so many Europeans and then Americans, Annie affirmed her steadfast belief in the value of friendship: "Friendship seems to me about the best thing this life gives us."[221]

On the other side of the continent, in California, girls began to attend Stanford University when it opened in 1891. Though it was a coed institution, Stanford nonetheless promoted same-sex friendships in numerous ways—for example, through sororities and other all-female associations. In the scrapbook assembled by Hazel Traphagen during her college years, she listed the names of all the Kappa Alpha Theta sorority members for the years 1902, 1903, and 1904—between a dozen and twenty names for each year. Presumably Hazel made friends with all or at least some of them. In addition, she could meet potential friends in the various clubs she labeled as "feminine," among them the Women's Athletic Association, Roble Gymnasium Club, Girls' Mandolin Club, and Girls' Glee Club. Some clubs, such as geology and band, were listed as "masculine," while others, such as orchestra, library, golf, and science, were designated as "neuter." With so many same-sex and coed clubs at her disposal, Hazel Traphagen had ample opportunity to meet other students with similar interests outside the hours taken up by course work.

Another Stanford scrapbook, that of Carrie Jette Johnson, from 1909 to 1911, indicates that she was a member of the Phi Delta Phi sorority and the Women's League. She attended receptions given by the YMCA and the YWCA, concerts and plays, and the wedding of her friend Clara Black. Among the many mementos distributed throughout her scrapbook, one literally stands out: a smudgy envelope dated May 9 to 12, 1909, bearing the words "envelope containing Clara Black's wedding cake which I slept on to determine my fate." (According-ing to an old superstition, if you sleep with a piece of wedding cake under your pillow you will see in your dreams a vision of your future spouse.) Girls' friendships, then as now, often ran parallel to heterosexual relationships and were expected to prop them up. If you couldn't tell your best friend the latest details of your boyfriend's daring proposals, as well as his shortcomings, in whom else could you confide?[222]

Even in a co-ed institution like Stanford, many academic programs were heavily sex-segregated. From the Stanford commencement programs of 1910 and 1912, one sees that men had a monopoly on engineering, geology, mining, and law degrees, whereas women were sprinkled liberally across the disciplines, with heavy concentrations in English, foreign languages, and history.

## Work Versus Marriage

The first generation of New Women, who came of age between 1880 and 1900, were aware that their young-adult freedom would probably end with marriage. That a stark choice would have to be made between marriage-and-family and a real career was a societal given, even for educated women. As the scientist Alice Hamilton put it in 1890: "The proper state of society is one in which a woman is free to choose between an independent life of celibacy or a life given up to childbearing and rearing the coming generation."[223] Those who chose a career looked to their female friends to create a supportive circle and ersatz family. Some entered into long-lasting domestic partnerships with other women.

Typically, women college graduates worked for a few years before getting married. Women with graduate or professional degrees usually had to commit to being permanently single, because employment at their level was considered incompatible with marriage and a family. For example, the fledgling professor Frances Rousmaniere had to give up her teaching post at Mount Holyoke when she married, as did Alice Freeman, the president of Wellesley College, when she became the wife of a Harvard professor.

Elizabeth Cady Stanton's granddaughter Nora Stanton Blatch recalled facing this dilemma at Cornell at the beginning of the century: "We all determined to combine marriage and careers, somehow. It would be hard, but it could be done."[224] But, alas, most women found that it could not be done, especially after children came along. Few middle-class women had the fortitude—or the desire—to follow the example of Nora's grandmother, who had decided in midlife to put work (albeit unpaid) and friendship at the center of her existence.

### Working-Class Friendships

Working-class women often did not have the choice to remain unemployed, even after marriage, especially if they were immigrants. From 1880 through 1919, twenty-three million immigrants came to the United States, mainly from Germany, Scandinavia, Ireland, Italy, Russia, Poland, and Romania but also from Canada and Latin America. In addition, the West Coast experienced a steady flood of workers from China and Japan. Overall, the population swelled by an estimated 25 percent. Immigrants streamed into Eastern and Midwestern cities, especially New York and Chicago, which had already expanded as America changed from a rural to a majority-urban nation.[225] The number of Americans living in cities increased from ten million in 1870 to fifty-four million in 1920.

Whether they came from abroad or from rural America, working-class women sought employment in the hope that they could eventually rise above their low socioeconomic status. Yet underpaid jobs in crowded factories and unhealthy sweatshops were the lot of most female immigrants and poor whites. Black women, denied even factory employment, were relegated to cleaning and cooking in private homes. Middle-class white women with some education could find work as salesladies in department stores or clerical workers in offices.

The better-educated daughters of the middle and upper classes filled the ranks of teachers, journalists, the new position of social workers, and the few slots open to them in medicine and law.

All these women needed friends to share their hardships and dreams as they struggled to find their way in an unfamiliar world. For the most part, friendships were made between members of the same social class. While Irene Tracy from the Good Will Club of New York City could optimistically write that her club provided the opportunity for a working-class girl like herself to befriend a woman from the leisured class, such friendships were highly exceptional. The club sponsors rarely, if ever, invited working-class women into their homes. As one astute observer expressed it, "Sisterhood might exist, but rarely across racial and class barriers."[226]

A Jewish immigrant from Poland, Sadie Frowne, told a typical story of a friendship formed with a girl from her own working-class milieu. Around 1900, Sadie was working ten hours a day, six days a week in a New York sweatshop making skirts, for which she earned four dollars a week. She lived with a girl named Ella, who worked in the same factory and earned five dollars a week, probably because she had more experience. In Sadie's own words:

> We had a room all to ourselves, paying $1.50 a week for it, and doing light housework . . . We did our cooking on an oil stove, and lived well, as this list of our expenses for one week will show.

## ELLA AND SADIE FOR FOOD (ONE WEEK)

Tea $0.06, Cocoa .10, Bread and rolls .40, Canned veg-
etables .20, Potatoes .10, Milk .21, Fruit .20, Butter .15,
Meat .60, Fish .15, Laundry .25.

The weekly bill for food and rent added up to a grand total
of $3.92—that is, less than half their combined salaries of
$9. Sadie was proud to add: "Of course, we could have lived
cheaper, but we are both fond of good things and felt we could
afford them." Of her spare money, Sadie spent one dollar a
week on "clothing and pleasure" and saved the other dollar.

This picture of female bonding born of economic necessity
was a far cry from the romantic outpourings or lofty sentiments
of young women from the better classes. Though Sadie did not
elaborate on her feelings for her roommate, she suggested that
they got along well, shared an appreciation for good food, and
did not deny themselves inexpensive pleasures. On their day off,
New York working-class girls like Sadie and Ella would often
go with their friends—both men and women—to amusement
parks, such as Luna Park in Brooklyn's Coney Island. They also
went to the theater and, by 1910, to one of the four hundred
moving-picture venues that existed in New York City.

Sadie would go on to attend night school, change jobs
and residences, increase her earnings to $4.50, and join a
union. In time she would acquire a boyfriend who urged her
to marry him, but—at the tender age of seventeen, when she
wrote down her story—she felt she was not yet ready.[227]

Working-class girls like Sadie often lived outside a family setting. In the wake of the Triangle Shirtwaist Factory fire of 1912, in which 146 garment workers lost their lives, the American Red Cross found that a third of the 123 female victims had lived alone or with roommates and subsisted entirely upon their own earnings.[228] Throngs of women like these constituted a new breed of unmarried workers, less subordinate to their families and more dependent on friendships with other women to make ends meet and share household responsibilities.

After they married and had children, working-class women were sometimes able to leave paid employment and stay home to take care of their families. Generally, they did not venture far from the tenements where they lived. While men would meet regularly after work in saloons, fraternal orders, or unions, women spent their evenings getting children to bed and cleaning up. Whatever leisure time they had was spent on their doorsteps with other women from their neighborhood, in local parks with their children, and in churches or synagogues. This didn't mean that they were bereft of friends. As in the early days of American history, neighbors might stop by at any time of day to borrow a kitchen item or ask for help in an emergency. This is demonstrated in a description of two Irish families living in an Upper East Side tenement: "Mrs. H. is very often in the house of Mrs. C., and they exchange many favors in the course of a day, while at night their husbands play cards and share the beer."[229]

Cities also provided new public spaces, such as parks, department stores, and movies, where people who scarcely knew one another could meet. These spaces offered opportunities for young people in particular, though women had to be wary of making the wrong kind of acquaintance—someone who could draw her away from the respectable values of her family and religion or, worse yet, cause her to lose her virginity and get pregnant. Seduction and rape were not just the bugaboos of pulp fiction; they were realities that working-class women encountered more frequently than did their more advantaged counterparts. When in distress, women turned to their friends for empathetic counsel, sometimes depending on them for advice about contraception and even abortion. Word of mouth between friends was the main way women obtained such information before 1916, when Margaret Sanger opened the first birth control clinic in Brooklyn and began to promote family planning nationwide.

Fortunately, some early progressive forces were sensitive to many other needs of the urban poor and provided new services such as free kindergartens, gymnasiums, swimming pools, adult education, and summer camps. These services, initially supported by women's clubs, were to find their fullest expression in the settlement house movement inaugurated in the United States by Jane Addams.

### Groundbreaking New Women: Jane Addams, Ellen Gates Starr, and Mary Rozet Smith

Jane Addams (1860–1935) cofounded the historic Hull House in Chicago with her close friend Ellen Gates Starr (1859–1940). They lived at Hull House in its early years and created the model for other American settlement houses, where women and men lived, worked, and agitated for social change. Later Addams began a relationship that would last more than thirty years with the rich Chicago heiress Mary Rozet Smith (1868–1934). None of these women ever married, and it is clear that their primary attachments were to other women. What interests us here is how their friendships played into the groundbreaking work they did at Hull House for more than half a century.

No one could possibly have imagined that the young Jane Addams would become a towering figure in American history, the founder of the first major American settlement house, a galvanizing force for social and moral justice, and a recipient of the Nobel Peace Prize in 1931. Born into a prosperous old American family in Cedarville, Illinois, she was stricken with polio at the age of four and subsequently suffered from curvature of the spine and other health problems. Later, a spinal operation, a rest cure, a nervous breakdown, and several episodes of depression left her unable to follow the career she had hoped for in medicine. Yet the early moral education she received from her father, her teachers at the Rockford Female Seminary in Illinois, and the works of Dickens, Tolstoy, and John Stuart Mill inspired her to consider becoming a social reformer. When she read in a magazine a description of Toynbee Hall in London, the first English settlement house, she was determined to see for herself what it had to offer.

In December 1887, with her Rockford Seminary friend Ellen Starr and several other friends, she set off for Europe, anticipating her visit to Toynbee Hall and dreaming that she might someday create something similar in America. It was while traveling in Spain—even before she had actually seen Toynbee Hall—that Addams confided her dream to Ellen Starr. This is how she remembered that conversation in her book, *Twenty Years at Hull-House*, published in 1910: "I can well recall the stumbling and uncertainty with which I finally set it forth to Miss Starr, my old-time school friend, who was one of our party." It was only with "the comfort of Miss Starr's companionship," and her vigorous enthusiasm, that Addams began to believe in the validity of her project.

After she and Ellen Starr parted from each other in Paris, Addams journeyed alone to London and finally saw Toynbee Hall, which exceeded her expectations. Its model of university men living among poor people and the mutual benefits they derived from such an arrangement would directly inspire the establishment of Hull House.

Back in the United States, Addams and Starr set about looking for a suitable building for their project. Addams described their luck in finding the old Hull mansion:

> The next January found Miss Starr and myself in Chicago, searching for a neighborhood in which we might put our plans into execution . . . The house had passed through many changes since it had been built in 1856 for the homestead of one of Chicago's pioneer citizens, Mr. Charles J. Hull, and although battered by its vicissitudes, was essentially sound . . . We furnished the house as we

would have furnished it were it in another part of the city, with the photographs and other impedimenta we had collected in Europe, and with a few bits of family mahogany . . . Probably no young matron ever placed her own things in her own house with more pleasure than that with which we first furnished Hull-House.

In 1889, once the house had been made suitable for visitors, Starr started a "reading party." At the first meeting, she read aloud from George Eliot's historical novel *Romola* to a group of young women, who then returned for subsequent weekly events held in the upstairs dining room. Two members of the group were invited to dinner each week, to be entertained as guests and also to help Starr and Addams afterward with the dish washing.

Starr also set in place other forms of cultural instruction, including an art course that she herself taught. Later, Starr's interest in the arts took another direction. As Addams recalled, Starr trained as a bookbinder and then established a bindery at Hull House, "in which design and workmanship, beauty and thoroughness are taught to a small number of apprentices."

Hull House would eventually become the residence of about twenty-five women. It would hold classes for hundreds of daytime visitors, many of which were taught by volunteers. It had a girls' club, a boys' club, a coed drama society, a kindergarten, a gym, a bathhouse, a library, a music school, an art gallery, and adult evening classes. By 1910, Hull House had become a thirteen-building settlement complex with a playground and a summer camp.

The program was especially attuned to the needs of poor children, many of whom left school at fourteen, after which attendance was no longer compulsory. Addams wrote: "It seems to us important that these children shall find themselves permanently attached to a House that offers them evening clubs and classes with their old companions, that merges as easily as possible the school life into the working life."

At the same time, Hull House provided special classes and services for adults and the elderly, which, according to Addams, dispelled the popular belief that "grown people would not respond to opportunities for education and social life." One example of their care of an old woman of ninety proved exceptionally rewarding:

> Left alone all day while her daughter cooked in a restaurant, [she] had formed such a persistent habit of picking the plaster off the walls that one landlord after another refused to have her for a tenant. It required but a few weeks' time to teach her to make large paper chains, and gradually she was content to do it all day long, and in the end took quite as much pleasure in adorning the walls as she had formerly taken in demolishing them . . . In course of time it was discovered that the old woman could speak Gaelic, and when one or two grave professors came to see her, the neighborhood was filled with pride that such a wonder lived in their midst.

Addams and Starr were often asked why they lived on Halsted Street when they could afford to live somewhere else. Their answer speaks to the heart of their shared philosophy: "Perhaps even in those first days we made a beginning toward that object which was afterwards stated in our charter: 'To provide a center for higher civic and social life; to institute and maintain educational and philanthropic enterprises, and to investigate and improve the conditions in the industrial districts of Chicago.'"

It is amazing to realize that two unmarried women without any special training were able to start such a haven for the underprivileged. Their undertaking was daring, not just for their times but for any age. Could either of them have done it alone? Although Addams was the leader of the project, would she have proceeded without the friendship and companionship of her own "Miss Starr"? Like earlier duos—Mechtilde of Hackeborn and Gertrude the Great, Sor Juana and María Luisa, Elizabeth Cady Stanton and Susan B. Anthony— Addams and Starr needed each other to find their best selves and realize their improbable dreams.

While Addams depended on Starr in the early years of Hull House, her most serious emotional attachment was to another woman—the wealthy and gracious Mary Rozet Smith. In her autobiography, Addams introduced Smith casually as witness to her long convalescence from typhoid fever during the autumn of 1895: "The illness was so prolonged that my health was most unsatisfactory during the following winter, and the next May I went abroad with my friend, Miss Smith, to effect if possible a more complete recovery."[230]

Their trip was monumental, to say the least. Not only did they visit many distinguished public figures in England but they also journeyed as far as Russia to meet with none other than the great Leo Tolstoy at his home, Yasnaya Polyana. Afterward they went to Germany, where they treated themselves to Wagner's *Ring* at Bayreuth—a far cry from Tolstoy's doctrine of agrarian simplicity. This spectacular journey initiated a partnership that kept Smith and Addams together for the rest of their lives.

Many in the lesbian community have claimed Addams and Smith as role models. The respected historian Blanche Wiesen Cook believed that they were lesbians, and a 2008 television documentary, *Out and Proud in Chicago*, also presented them as such.[231] Other scholars have taken the position that there is no way of knowing whether Addams and Smith had a sexual relationship, and that it is just as likely that they had a nonsexual romantic friendship.[232] However we define their relationship today, they called each other "friends" and lived up to the highest ideals of that meaningful word. For them it meant sharing their friendship with the distressed immigrants and downtrodden poor of Chicago's urban slums. It also meant traveling together, writing each other loving letters when they were apart, and thinking of their relationship as a kind of marriage. Mary Rozet Smith's portrait hangs in the Hull-House Museum, not only by virtue of her friendship with Addams but also because she was a major financial contributor to the upkeep of the house.

## Hilda Satt and Jane Addams

It is fascinating to view Jane Addams and Hull House from the perspective of a poor Jewish immigrant girl. Hilda Satt, living with her widowed mother and her sisters in a working-class neighborhood, was first taken to Hull House by an Irish playmate. They went together to a Christmas party, where children of various religions from various countries speaking various languages were all thoroughly enjoying themselves. Hilda was astonished at this vision of peaceful multiculturalism and equally astonished by the welcoming presence of Jane Addams. As Hilda recounted in her autobiography, "It was the first time that I looked into those kind, understanding eyes. There was a gleam of welcome in them that made me feel I was wanted."

In the years to come, those understanding eyes would follow Hilda into young adulthood, encourage her after she dropped out of school to become a garment worker, direct her into an adult-education English course, and even see to it that she spent a semester at the University of Chicago, though she had not earned a high school diploma. Eventually Addams found a place for Hilda as a paid worker at Hull House. Jane Addams became Hilda's patron, mentor, guiding spirit, and ultimately her friend. In Hilda's words: "For ten years I spent most of my evenings at Hull House. The first three years of that time I saw Jane Addams almost every night . . . her presence was always felt, whether she was there in person or in spirit."

Hilda went on to work in a publishing house and to marry a man from a middle-class background, which occasioned a move to Milwaukee. But after fifteen years of comfortable family life, when her husband died in 1927 and she lost most of their fortune, she left Milwaukee, moved with her children back to Chicago, and resumed her relationship with Jane Addams, now on a more equal footing.

At the fortieth anniversary of the founding of Hull House, Hilda Satt Polacheck sat at Jane Addams's table in the dining room, among many celebrities and former members of their community: "Jane Addams moved among the great and the humble just as any mother would when her far-flung children returned to the old home for a reunion. She knew everybody's name. She asked after children of the former children who had come to Hull-House years ago as bewildered, uprooted little immigrants." At that moment in Hilda's life, after the harrowing losses of her husband and money, she found in Jane Addams a supporting friend who gave her "a sort of transfusion of hope and courage that helped through the dark days that lay ahead."[233]

Granted, Hilda's picture is colored by her gratitude to her former mentor and her idolatrous vision of Addams as the patron saint of Hull House. Nonetheless, the bonds of friendship encompassed these two women of unequal stature and endured until Addams's death in 1935.

*Education, Class, and Race as Factors in Friendship*
Around 1900, a new model of female friendship emerged that was to last for much of the twentieth century. Among the middle and upper classes, friendships were predicated on a cer-

tain level of education—at least high school and often college. One's school, as much as or perhaps more than one's church, provided the setting for finding and maintaining friendships. Education was the means whereby the daughter of Greek immigrants could become the college friend of a debutante with roots in colonial America.

During the first decades of the twentieth century, working-class women were lucky if they were able to finish high school. Most did not. Leaving school at sixteen or even fourteen meant that they would make friends only of their own educational level and social class, though a few—like Hilda Satt—moved into the middle class by virtue of adult education or an advantageous marriage. As for women of color, whether they penetrated the middle class or not, the racial divide prevented them from making friends with white women, especially in the South. Legal segregation in Southern public schools did not end until 1954, and de facto segregation would last much longer.

Middle-class white women in the South continued to meet with one another regularly to enjoy "bridge, tea, and gossip," according to the observation of one woman scholar in 1925. Still, even in the South, after World War I a progressive spirit animated many women's communities. Some Southern women—like women in other parts of the nation—gathered with their peers not only for social occasions but also for serious efforts to ameliorate the condition of disadvantaged children, women workers, and African Americans.[234]

Sometime during the 1920s the public image of the serious New Woman began to fade in America, to be replaced by that of the carefree flapper. Just as Lucy Capehart in 1904 had looked back on nineteenth-century romantic friendships as a thing of the past, so, too, did Vida Scudder, a Wellesley professor, declare in the 1920s that the crusading comradeship she had experienced at the turn of the century was no longer operative in the lives of younger women.[235] The hard-won rights gained by one generation of social activists were taken for granted by their daughters and granddaughters.

Yet the changes wrought by such reformers as Elizabeth Cady Stanton, Susan B. Anthony, and Jane Addams did not disappear. New educational and employment opportunities made it possible for some women to live alone or, as we have seen, in domestic partnerships that provided an alternative to marriage. In her crusade to make contraception legal, Margaret Sanger made it possible for many women to limit the number of their children and to enjoy greater sexual freedom. Women's suffrage, passed in 1920, gave all adult women a voice, however muted, in government and public affairs. These determined reformers proved that women working together could address any number of social ills, and, contrary to the disillusion expressed by Vida Scudder, their examples perpetuated the belief that anything was possible for the daughters of America.

### Vera Brittain and Winifred Holtby

A striking portrait of friendship between two New Women was written not by an American but by an Englishwoman.

In her *Testament of Friendship,* Vera Brittain (1893–1970) wrote the story of her close companionship with Winifred Holtby (1898–1935) from the time of their meeting at Oxford until the moment of Holtby's premature death from kidney failure. Brittain and Holtby's friendship was remarkable not only for the deep affection between them, which continued long after Brittain had married, but also for the way each encouraged the other in her work as a writer and reformer. Brittain's testament reveals the qualities we have emphasized as characteristic of the best women's friendships throughout the ages—intimacy, self-disclosure, shared experiences, loyalty, nurturance, loving concern, and mutual support. At the same time, her testament manifests the distinctive features of friendship among educated British women immediately following World War I.

Both Brittain and Holtby returned to Somerville College at Oxford in 1919. Brittain, the elder of the two, was experiencing a severe postwar depression, having lost in the war her fiancé, her brother, and her brother's two closest friends. After four years of war service as a nurse and the excruciating pain of her losses, she was thoroughly exhausted and deeply pessimistic. Holtby, still fresh and optimistic after only one year with the WAAC (Women's Army Auxiliary Corps), would prove to be the high-spirited presence who would help bring Brittain back to life.

Women had been students at Oxford since 1878, but they became eligible to receive degrees only in 1920. Thus Brittain and Holtby were among the very first women granted an

undergraduate degree from one of the Oxford colleges designated for women, in their case Somerville. But even before
they shared that bond, Brittain and Holtby were, in Brittain's
words, "invisibly linked by the unique bond of our war service
and the attitude towards it of mingled pity and disapproval
which we sensed . . . on the part of the Somerville dons."

Initially, Brittain was mistrustful of Holtby's upbeat personality and expressed a barely concealed hostility toward
her. But the link of their war service was too strong to be
denied, and when they found each other in the cathedral on
the first anniversary of Armistice Day, they discovered that
they were kindred spirits. They quickly confessed their desire
to be writers and began the endless talks, walks, and travels
that would endure for the next sixteen years. When their first
year at Oxford ended, they took a fortnight trip together to
Cornwall, where they walked "through honeysuckled lanes or
along the rocky coast" and prepared their work for the following term, when they would live together in off-campus rooms.

They would talk of their lost loves—Brittain's fiancé killed
in the war, Holtby's wounded and estranged young man. They
would sit in front of the fire as Holtby, her face in Brittain's
lap, poured out the torment of her religious conflicts. Despite
the agony of shattered romances and religious doubts, they
were both certain that their dearest ambitions lay in becoming journalists and novelists. With their degrees in hand, they
set off for shared lodgings in London.

Their New Woman stance was articulated in a little verse
that Holtby wrote to a tune from *The Mikado*:

*We mean to run this show.*
*We are not shy.*
*We'll make the whole world go—*
*My friends and I!*

Belief in the power of friendship shored up their spirits and girded their projects as they left their studies behind and took on the world. Holtby began writing and lecturing on subjects she cared about fervently, such as world peace. When she was away from Brittain, she proved herself to be "a gay, grateful, infinitely responsive letter-writer, whose correspondence suggested a long, vivid, unbroken, conversation." Their greatest joint pleasure was a summer trip to Italy, which Brittain considered "the most perfect holiday of all my experience and, I believe, of Winifred's." After they parted, Winifred wrote to Vera a letter in which she expressed her deep affection: "The best thing of all was finding out from day to day how dear you are. The journey would have been pleasant in most circumstances, . . . but because you were there, it was wholly delightful." Yes, friendship can be encapsulated in the unique pleasure one experiences with a specific person: because you were you, and I was I. In this respect, true friendship is not so different from true love.

Between 1922 and 1923 Brittain and Holtby shared flats in Bloomsbury, not far from the British Museum. Then they moved into a spacious flat in a mansion in Maida Vale, a less fashionable district than pricey, intellectual Bloomsbury. Brittain remembered their "exhilarating certainty" in the 1920s that "mankind could learn" not to repeat the horrors of World War I. But, tragically, they were wrong.

Each pursued her work as a journalist and fiction writer, with growing success. Each would return from visits to her family of origin "more than ever glad of our strenuous, independent, enthralling London existence." Like many happy couples, they had so much to tell each other after a day's work: "Neither of us had ever known any pleasure quite equal to the joy of coming home at the end of the day after a series of separate varied experiences, and each recounting those incidents to the other over late biscuits and tea."

The closeness established between the two women was such that they responded to each other's needs and emotions in what Brittain considered an instinctive fashion—that is, they answered a statement or request even before it had been made. People who have been together for a long time know exactly what Brittain means. Married couples, domestic partners, sisters, and mothers and daughters are sometimes apt to finish each other's sentences.

Yet Brittain and Holtby were not mirror images of each other. Theirs was a complementary friendship, in appearance as much as in temperament. Brittain was small, dark-haired, and troubled, while Holtby was tall, blond, and outgoing. Brittain was the elder, more damaged, and in some ways more needy of the two. Retrospectively, she was able to acknowledge that she had probably been very difficult to live with after the tragedies of the war, and that the survival of their friendship was primarily Winifred's achievement rather than her own.

Despite her precious friendship with Winifred, Vera Brittain accepted the proposal of a young university professor, George Catlin, and married him in 1925. The day before the wedding, Holtby gave Brittain a platinum chain studded with seed pearls, to be worn with her wedding dress. (When Holtby lay blind on her deathbed twelve years later, Brittain wrapped the chain like a bracelet around her own arm and closed Holtby's fingers over it.) Holtby was Brittain's only bridesmaid, beautifully attired in a large feathered hat that was complemented by an enormous bouquet of blue and mauve delphiniums. Only much later did Brittain acknowledge to herself that her marriage must have caused Holtby very painful emotions, which she had managed to conceal "with so loving and self-forgetful a magnanimity." Holtby recorded her feelings in a poem called "The Foolish Clocks," about the ticking of the clocks that Brittain had been accustomed to wind in their flat: "Now she is gone, but all her clocks are ticking." The clocks are a loud reminder of "the precious moments when my love was here."

Fortunately, the marriage did not break up the friendship. Each woman continued to support the other in her work as a novelist, journalist, and campaigner for the League of Nations. After Holtby's five-month visit to South Africa, she often lived in London with Brittain and Catlin as an honorary family member. In time, Brittain would have two children, a son and a daughter, whereas Holtby would remain unmarried.

Though Holtby lived only thirty-seven years, she managed in that short span to publish six novels, numerous short stories, poems, works of satire, and a book on Virginia Woolf. She gave speeches on pacifism and the rights of women and Africans, and had become a notable public figure by the time of her death.

Brittain's first novel was published in 1923, but she did not find her true literary voice until she wrote her autobiographical works, *Testament of Youth* (published in 1933) and *Testament of Friendship* (1940). The latter was essentially a biography of Winifred Holtby, with a large portion devoted to their sixteen-year friendship. As an author, Brittain was fully aware that friendships between women were rarely part of the historical record:

> From the days of Homer the friendships of men have enjoyed glory and acclamation, but the friendships of women in spite of Ruth and Naomi, have usually been not merely unsung, but mocked, belittled and falsely interpreted. I hope that Winifred's story may do something to destroy these tarnished interpretations, and show its readers that loyalty and affection between women is a noble relationship which, far from impoverishing, actually enhances the love of a girl for her lover, of a wife for her husband, of a mother for her children.[236]

We would add only that women today do not need to justify their friendships by presenting them as positive adjuncts to heterosexual love, marriage, and motherhood. But in Brittain's era, her intense relationship with Holtby was considered unusual and even suspect. To our eyes, those two New Women, united by personal affection and a shared commitment to meaningful work, paved the way for later twentieth-century women to rally together under the banner of sisterhood.

TEN

# ELEANOR ROOSEVELT
# AND HER FRIENDS

*"We drank a toast to absent friends whom we would like to have with us at dinner & I thought of you dear one as I proposed it."*

—LETTER FROM ELEANOR ROOSEVELT TO LORENA HICKOK,
CHRISTMAS NIGHT IN THE WHITE HOUSE, 1933

*"Our friendship is such an old one that time and distance
have never made any difference."*

—LETTER FROM ELEANOR ROOSEVELT TO ISABELLA SELMES
FERGUSON GREENWAY KING, OCTOBER 9, 1953

*"There is no more precious experience in life than friendship."*

—*ELEANOR ROOSEVELT'S BOOK OF COMMON SENSE ETIQUETTE*, C. 1962

ELEANOR ROOSEVELT (1884–1962) LIVED LARGE at the fulcrum of social and political change that affected all Americans, women in particular. She personally experienced the radical transformations in women's roles that stretched from the Victorian ideal of the angel in the house to the New Woman at the turn of the century, the even Newer Woman of the 1920s and 1930s, and further into wartime disruptions, postwar optimism, the Civil Rights Movement, and the early stirrings of organized feminism. And throughout Eleanor's remarkable life, her many friends acted as midwives to her psychological, social, and political rebirths.

Though they were long eclipsed by Franklin and Eleanor's public personae, Eleanor's friends have by now become familiar figures to historians of the Roosevelt years. Unlike other American presidents' wives, with the exception of Abigail Adams, Eleanor left behind a mound of written material that attests to her extensive network of family members and friends. This well-examined trove of letters, diaries, autobiographies, articles, and speeches yields rich insights into her personal relationships and spotlights the overall scene of women as friends in the first half of the twentieth century.

From the time that Franklin Roosevelt became governor of New York in 1928 until his death in 1945, Eleanor was seen—and took care that she was seen—primarily as his devoted helpmate. Franklin's fireside chats on the radio and Eleanor's "My Day" newspaper column, plus constant media attention to their every move, gave them star billing hitherto unknown to any presidential couple. They projected the image of a harmonious, compassionate, patrician pair committed to the welfare of everyday

Americans. Few people knew that he and his wife had long ceased sharing the marital bed and that they both had sustaining friendships outside the marriage.

Married at twenty in 1905, Eleanor gave birth to six children during the first ten years of her marriage and was shocked to discover, in 1918, that Franklin had been having an affair for two years with her own social secretary, Lucy Mercer. Eleanor offered him a divorce, but Sara Roosevelt—Franklin's imperious mother, with whom they lived for much of the year at her Hyde Park estate—threatened to cut off financial support if he accepted. Eleanor and Franklin remained married, and she proved to be an indispensable ally in his political career, especially after he was stricken with polio in 1921 and left paralyzed from the waist down.

In the ensuing years, Eleanor turned increasingly to her friends for love and validation. They helped her develop beyond the insecure young woman who had been traumatized by the death of her mother when she was eight and then that of her adored father (the younger brother of President Theodore Roosevelt) when she was ten. Born into old New York society, raised by her strict maternal grandmother, and sent to finishing school in England, Eleanor returned to New York for her debut in 1902, when she was eighteen. Although she subscribed to the rigidities and pieties incumbent upon a young debutante, she also took up volunteer work at the Rivington Street settlement house on the Lower East Side, which gave voice to her compassionate nature. Franklin Roosevelt, a distant cousin and Harvard student, would draw her in another direction—into the supportive role of wife to a man with political ambitions shared by his mother.

*Isabella Selmes Ferguson Greenway King*

At the time of her marriage, Eleanor was already friends with the classically beautiful Isabella Selmes. Though two years younger, Isabella became and remained a constant figure in Eleanor's life. In fact, when Franklin proposed, Eleanor confided their secret engagement to Isabella even before the young couple told their families. Isabella was one of Eleanor's six bridesmaids. Within months of Eleanor's wedding, Isabella married Robert Ferguson, a man eighteen years her senior and a longtime friend of the Roosevelt family. The two couples visited with one another on Robert's ancestral estate in Scotland during their respective European honeymoons. In New York City, where Eleanor and Isabella lived in similar brownstones, they exchanged advice about childbearing, child raising, and household management. Isabella was godmother to the Roosevelts' first child, Anna, born in May 1906, and Eleanor was godmother to Isabella's first child, Martha, born in September of the same year. Each daughter was followed by a son—James Roosevelt in 1907 and Robert Ferguson in 1908. Even with young children, Eleanor and Isabella met often at the Three Arts Club, which provided lodging for women in the arts, or at the Junior League, founded for the promotion of settlement houses.

But this idyllic upper-class life came to an abrupt end when Bob Ferguson was diagnosed with tuberculosis and the Ferguson family was forced to retreat from New York City to a more forgiving climate in New Mexico. Eleanor would write frequently to bolster Isabella's spirits, but she, too, would undergo a tragic ordeal: the loss of her third child, Franklin, at seven months from influenza on November 1, 1909. Her letter to Isabella dated November 12 expressed the comfort she received from her friend's courageous example: "I wanted to tell you dear that your unselfish cheerful example in all your anxiety & sorrow has helped me these past weeks more than you will ever know . . . Sometimes I think I cannot bear the heartache which one little life has left behind but then I realize that we have much to be grateful for still."

After the Fergusons moved west, Eleanor and Isabella's steady correspondence over more than four decades conveyed the deep affection they felt for each other. Their letters contained stories of Eleanor's new babies—Elliott in 1910, Franklin D. Jr. (named in memory of the baby she had lost) in 1914, and John in 1916—and both women's social and political activities. Occasionally they were even able to see each other face-to-face, both on the East Coast and in the West.

When the United States declared war on Germany in April 1917, the two women undertook war-related activities, Eleanor with the Red Cross in Washington, DC, and Isabella with the Council for Defense in New Mexico. As was the case for Vera Brittain in England and other New Women, the war provided an opportunity for Eleanor and Isabella to take on meaningful volunteer work, an important step in their future lives as public figures.

In the fall of 1922 Bob Ferguson died of kidney failure, and Isabella wired Eleanor at once. Eleanor responded with a heartfelt letter expressing her love for Bob, whom she considered "a devoted, loyal friend." When Isabella mustered up the strength to write back three months later, she began her letter with an acknowledgment of their unbreakable bond: "Oh my beloved Eleanor—With you one doesn't begin—One just goes on—and *you* are the wonderful friend who has made that possible in the face of my inexplicable *silence*." She knew that Eleanor would understand the "anguish and mortification" that had kept her from writing until then. And she knew that Eleanor would approve of her marriage less than two years later to John Greenway, with whom she moved to Arizona.

Isabella's marital happiness was short-lived. In early January, 1926, she accompanied her husband to New York for a gallbladder operation, and one week later, he died in her arms. Eleanor assured Isabella that she would always be there for her: "If ever you & Martha [her daughter] really want me I can come for I love you oh! so dearly & long to help even though I know no one can."

In the following years, like many New and Newer Women, both Eleanor and Isabella were active in politics, each with the national Democratic Party. After Franklin's two terms as governor of New York state, he was nominated for president at the 1932 Democratic Convention in Chicago, and it was none other than Isabella Greenway—a delegate from Arizona—who made the first seconding speech for his nomination. And in August of the same year, Isabella won the Arizona Democratic primary over two male opponents and was elected to the seventy-third Congress to complete the term of a resigning representative. She moved to Washington for the start of her term, where she could meet regularly with her old friend, now first lady of the nation.

Sometimes Isabella lunched or dined at the White House. Sometimes Eleanor walked her dog over to the Willard Hotel to meet Isabella for lunch or tea. Isabella won reelection in 1934 but declined to run again in 1936. She married for a third time in April 1939, to Harry O. King, and spent her remaining years in New York and Tucson.

During Franklin's first and second terms in office, he had no greater supporter than Isabella Selmes Ferguson Greenway, but once he decided to run for a third term, Isabella decided to back Wendell Willkie instead of Roosevelt. Eleanor was, of course, disappointed, but she assured Isabella that her decision would not impact their friendship. She wrote: "I realize that you have to do whatever you think is right."[237] Franklin won reelection and the friendship continued as before.

When World War II erupted for Americans in 1941, Eleanor and Isabella again became involved in various forms of war work. This did not stop them from remembering each other's birthdays and sending gifts. They met in New York City on February 15, 1945, for what Eleanor hoped would be a really quiet hour. Whatever peace of mind Eleanor achieved was shattered a few months later by the news of Franklin's death on April 12, 1945, in Warm Springs, Georgia. Soon afterward, meeting with Isabella for lunch at the Biltmore Hotel in New York, Eleanor was trying to adjust to widowhood and to life outside the White House, her home for the past twelve years. It must have been a great comfort to be with her lifetime friend, from whom she had no secrets.

### New Friends, New Worlds

There had been secrets, indeed, in Eleanor's life. Recounting the story of her unhappy childhood and the discovery of Franklin's affair with Lucy Mercer became a rite of passage to mark the transition from mere acquaintanceship to real friendship. Eleanor's new friends in the 1920s and 1930s included Esther Lape and Elizabeth Read, Nancy Cook (Nan) and Marion Dickerman, Elinor Morgenthau, Malvina Thompson (Tommy), Louis Howe, Earl Miller, and Lorena Hickok. Each of these friends was instrumental in helping Eleanor step outside her family circle and engage in the greater world, where she would make her mark as a spokesperson for liberal causes.

In Greenwich Village during the 1920s a number of New Women lived singly, or in pairs that were still known as Boston marriages. These women had campaigned for women's suffrage, fought for the abolition of child labor and better conditions for working women, and participated in other progressive endeavors. Among them, Esther Lape and her partner, Elizabeth Read, and Nancy Cook and her lifelong companion, Marion Dickerman, would serve as mentors to Eleanor in her political apprenticeship.

Eleanor came to know these independent women first through her work for the League of Women Voters and the

World Court, and then through purely social evenings spent together in their Greenwich Village apartments. With Esther Lape, a Wellesley graduate and college professor, and Elizabeth Read, a Smith graduate and attorney, Eleanor shed the cloak of the aggrieved wife and dutiful daughter-in-law and became one of the girls, happy to dine informally and read poetry out loud.[238]

Another 1920s Greenwich Village couple, Marion Dickerman and Nancy Cook, figured prominently in Eleanor's life for two decades. Nan was the director of the Women's Division of the New York State Democratic Party, and Marion was a teacher and vice principal of the Todhunter School for Girls in New York City. They helped Eleanor set up women's Democratic clubs throughout upstate New York, and together they founded a newsletter called *Women's Democratic News*. Considering Eleanor's friendships with Marion, Nancy, Esther, and Elizabeth, the historian Doris Kearns Goodwin has written: "There is every evidence that the four women, along with a half a dozen others . . . played a substantial role in the education of Eleanor Roosevelt, tutoring her in politics, strategy, and public policy, encouraging her to open up emotionally, building her sense of confidence and self-esteem."[239]

In Hyde Park, Eleanor was obliged to live according to her mother-in-law's posh standards, but only two miles away from the estate she could retreat with Franklin to his Val-Kill country property. Here, picnicking with Marion, Nancy, and Franklin, Eleanor enthusiastically accepted his suggestion that they build a small house there in which she could live more simply. The charming fieldstone cottage they constructed would become the first residence Eleanor considered truly her own. Here she could invite her friends home without asking permission from "Mama," who was known to dislike Eleanor's independent-minded companions.

Close friendships provided Eleanor with the ballast she needed to keep her marriage afloat. Confronted with Franklin and his vivacious personal secretary, Marguerite LeHand, known as Missy, and the circle of adulators surrounding her charismatic husband, Eleanor developed her own coterie of loyal women and men. None were more loyal in the 1920s than Marion and Nancy, who became her partners in Val-Kill Industries, which Eleanor created to revive local handicrafts. Under Nancy's direction, Val-Kill Industries became a successful venture in furniture making, weaving, and metalwork. Franklin encouraged Eleanor's three-way working relationship with Nan and Marion, and he was proud to place the first pieces of furniture produced at Val-Kill in his cottage in Warm Springs, Georgia. Warm Springs, the primary site for polio treatment in America, provided Franklin with a healthful retreat supervised by his personal secretary, Missy.

Eleanor's relationship with Missy was complicated. Since Franklin and Eleanor had agreed to lead separate private lives in what proved to be a successful, albeit unconventional, partnership, she could hardly object to his having a secretary-companion, especially since Missy freed Eleanor from having to pay constant attention to her paralyzed husband. Eleanor also knew that she—Franklin's wife, co-parent, and political ally—was indispensable to him, and she believed their private lives could be kept separate from the important work they wished to accomplish in the public sphere.

In addition to Val-Kill, Eleanor, Nancy, and Marion bought the Todhunter School for Girls in New York City. Marion became its principal, and Eleanor its most beloved teacher of literature and American history. From 1928 to 1932, while her husband was governor of New York, Eleanor would spend the first two and a half days of the school week teaching and then would rush back to Albany to take up her responsibilities as the governor's wife. Eleanor loved teaching and gave it up, reluctantly, only when Franklin became president. She considered her shared work with Nancy at Val-Kill and Marion at Todhunter as "one of the most satisfactory ways of making and keeping friends."[240] In this respect, Eleanor followed the path of Jane Addams and other turn-of-the-century New Women. The even Newer Women of the 1920s and 1930s, with whom Eleanor associated and whose careers she helped advance, would become role models for generations to come. It was no accident that President Roosevelt's official entourage included the first woman appointed to the US cabinet, Secretary of Labor Frances Perkins, who remained in that post throughout his entire tenure as president.

The Roosevelt administration promoted unprecedented numbers of women to senior government posts. Many of these appointees relied on friendship with the first lady for access to the president. This group understood one another's travails and were, for the most part, supportive. When FDR's death was announced to his cabinet, Frances Perkins and Eleanor "sat on a bench like two schoolgirls" and cried.[241]

Elinor and Henry Morgenthau, Jr. were the Roosevelts' first Jewish friends. From the start, Eleanor admired—and envied—Elinor's fine Vassar education. She fully appreciated Elinor's ability to run a smooth household with three children and a large staff, alongside her work in the Women's Division of the Democratic Party and her support for her husband, Henry, then the leading agriculturalist in the state of New York.

Yet sometimes, even with one's closest friends, there are moments of strain and misunderstanding. This was the case between Eleanor and Elinor in 1928 during Franklin's successful campaign for governor of New York, when Elinor felt she had been slighted. In an effort to assuage her friend's feelings but also to express her own misgivings, Eleanor wrote: "I have always felt that you were hurt often by imaginary things & have wanted to protect you but if one is to have a healthy, normal relationship I realize it must be on some kind of equal basis, you simply cannot be so easily hurt, life is too short to cope with it!"[242] And she invited Elinor to lunch that week, in the expectation that her friend would get over it. She did.

Later, when Henry Morgenthau, Jr. became President Roosevelt's secretary of the treasury, Elinor and Eleanor took to horseback riding before breakfast almost daily in Washington's Rock Creek Park. They would also go to New York together for theater and dinner at the Colony Club. It came as a shock to Eleanor when her friend was turned down for mem-

bership in the club because she was Jewish, whereupon Eleanor resigned in protest.[243] Because of her friendship with the
Morgenthaus, Eleanor was forced to confront the anti-Semitic
prejudices rampant in her class.

Another important friend was Malvina Thompson, called
Tommy, who started helping Eleanor with state politics in
the mid-twenties and remained her right hand for the next
thirty years. As Eleanor's secretary during her residence in the
White House, Tommy was always at Eleanor's side, whether it
was a matter of securing a nearby apartment for guests or ac-
companying Eleanor to London in the midst of World War II.

Tommy also became friends with Eleanor's daughter,
Anna, and it is to their extensive correspondence that we owe
another view of Eleanor's personal relationships. Tommy's letters to Anna are crucial for understanding Eleanor's painful
break with Marion Dickerman and Nancy Cook in 1938 to
1939. From Tommy's vantage point, Nan and Marion were
disdainful of the new people Eleanor brought to Val-Kill in
the 1930s—people from all walks of life, including left-wing
students and the children of immigrants and sharecroppers.
Nan and Marion resented the fact that Tommy worked and
slept at Val-Kill and often treated her cavalierly, as indicated
by Tommy's letter to Anna of September 10, 1937: "I must
tell you that Miss Dickerman has undertaken to complete my
education—she tells me I talk too loud—I use certain phrases
too often and emphasize words when I shouldn't! I'm having a
rare opportunity to polish off the rough corners!" In the end,
Eleanor bought out Nan and Marion's share in Val-Kill and

withdrew from the Todhunter School. The break was painful for all concerned, but especially for Eleanor, who had never expected to lose the friendship of women once so dear to her. Tommy's appraisal of Eleanor's erstwhile friends was cynical: "I felt so sorry for ER I could have wept and I don't think tears were very far away from her. I can't for the life of me understand why such a fine person as she is has so many chiselers around her."[244]

### Louis and Earl

Throughout this period Eleanor also had several close male friends, namely Louis Howe and Earl Miller, and, later, Joseph P. Lash. Louis Howe was Franklin's primary political adviser; Eleanor initially mistrusted him, yet she was ultimately able to see through his politically slick comportment to the kernel of gold inside. During the 1920s they developed a deep friendship that was to last until his death in 1936. Louis was the only one of Eleanor's friends who had an equally close relationship with Franklin, and, as such, he was able to play a unique role as a bridge between them. He brought her into politics by asking her to review speech drafts and discuss new ideas. She was flattered, and, under his guidance, she began to have more confidence in her contributions to Franklin's campaigns.

Earl Miller presents an entirely different story. He was Eleanor's bodyguard during Franklin's governorship, traveling with her night and day and protecting her from harm. Unlike Louis, he was tall, handsome, athletic, and something of a ladies' man. Though he was demonstrative with Eleanor in public, placing his arm around her shoulders or his hand on her knee, his intent seems to have been respectfully chivalrous. He always referred to her as "the lady." Eleanor's friends did not like the closeness that developed between the governor's wife and her plebeian bodyguard. Yet they recognized that she was more relaxed and playful when he was around, and that their affection was mutual. Marion Dickerman recalled: "He gave something to Eleanor . . . It was a very deep attachment."[245]

### *Hick*

Eleanor's friendship with Lorena Hickok, called Hick, has proven to be even more controversial. Their correspondence, from the early 1930s until 1962, consists of around 3,500 letters, which are archived in the Franklin Delano Roosevelt Presidential Library. The letters are a gold mine for anyone seeking to understand and portray the mature Eleanor Roosevelt. When the correspondence was opened to the public in 1978, it became clear that Eleanor had loved Lorena deeply, and, for a time, more passionately than any of her other friends.

In the fall of 1932, Hick was an Associated Press reporter assigned to cover Franklin's first presidential campaign. Nearing forty, she was a highly successful journalist in a man's field, known to drink and smoke with the best of them. A solid, tall woman weighing two hundred pounds, she fell short of Eleanor's even taller patrician figure.

Their upbringings and social positions could not have been

more different. Hick had grown up in South Dakota under the hand of a violent, abusive, working-class father. After her mother's death when she was fourteen, Hick had mainly been on her own for whatever schooling she achieved. She worked her way up in journalism through several newspapers before becoming a member of the Associated Press. Eleanor sympathized with Hick's horrific beginnings and admired the accomplished reporter she had become. Hick sympathized with Eleanor's fear of being swallowed up in Franklin's bid for the presidency and of losing any independent life of her own.

By the time Franklin was sworn in as president on March 4, 1933, Hick had become indispensable to Eleanor. From the White House, Eleanor wrote her lengthy letters almost daily. To Hick on March 7, Hick's birthday: "Hick darling, All day I've thought of you and another birthday I *will* be with you, & yet to-night you sounded so far away & formal, oh! I want to put my arms around you. I ache to hold you close. Your ring is a great comfort, look at it & think she does love me, or I wouldn't be wearing it!" To Hick on March 8: "Just telephoned you, oh! it is good to hear your voice, when it sounds right no one can make me so happy!" To Hick on March 9: "My pictures are nearly all up & I have you in my sitting room where I can look at you most of my waking hours! I can't kiss you so I kiss your *picture* good-night & good morning! "[246]

Reading these letters, we are thrown back to nineteenth-century romantic friendships. The language and sentiments are identical, with the exception of references to the telephone.

Expressions of affection, the ring Hick gave Eleanor, the pictures Eleanor displayed of Hick, the desire to embrace the other and to be reassured that one is loved in return—these are all present in Eleanor's new relationship.

In response to Hick's suggestion that Eleanor's life as first lady would be of interest to future biographers, she added her daily schedule at the end of each letter. Sent dutifully as a regular part of her correspondence, Eleanor's diary attests to her enormous vitality, which impressed Hick from the time she had covered Eleanor on the campaign trail and called her a "whirlwind." As first lady, Eleanor gracefully intermingled seeing family members and friends with performing her political duties, forming a fabric of social life that would surround her throughout her husband's entire presidency. Though they appeared together at political events, dinners, movies, concerts, and all other occasions that required the appearance of a united couple, Eleanor experienced real intimacy not with her husband but with her closest friends, and especially with Hick in 1933 and 1934.

She confided in Hick her worries about her children as their marriages ended in divorce. She sent running accounts of her travels from Los Angeles, which she reached by air; from Tucson, where she spent an evening with Isabella Greenway; and from New York City and Hyde Park. The good news was that Hick had taken a job in Washington at the Federal Emergency Relief Administration, with the assignment of gauging the effectiveness of the nation's New Deal programs. Though her job entailed considerable travel, she had the pleasure of returning to Washington, DC, and sleeping at the White House in a room adjacent to Eleanor's suite.

It was Hick who suggested to Eleanor that she hold weekly news conferences with the press, and that the reporters be limited to women. Eleanor was at first reluctant to subject herself to such publicity, but with Hick's encouragement—and Franklin's as well—she undertook what proved to be a highly successful public relations venture, and one that Eleanor, coached by Hick, even learned to enjoy.

In July 1933, Eleanor and Hick managed to take a road trip together to upstate New York, New England, and French Canada. Today it is hard to imagine the first lady of the United States driving her own car—and a blue Buick convertible at that!—without being recognized, or her traveling without the presence of security agents. Yet that's exactly what Eleanor did, for three delicious weeks. When they returned to the White House on July 28, Franklin quickly scheduled a private dinner so he could hear all about their adventures.

Eleanor and Hick sometimes found opportunities to work together on causes concerning the underprivileged. For instance, when Hick traveled to the coal mining region near Morgantown, West Virginia, she was so appalled by what she found that she telephoned Eleanor and asked her to come quickly. Eleanor, too, was struck by the dire poverty of the place, and she persuaded her husband to use that site for a subsistence homestead program—a project called Arthurdale that came to be known as Eleanor's Baby.

The letters exchanged between Hick and Eleanor in 1934

read like an emotional roller coaster. At times, the longing each felt for the other gushed forth in consecrated phrases: "I love you," "I miss you," "I love you dearly and very tenderly," "With all my heart I love you," "A World of love." At other times, the problems arising from their different life situations and temperaments interfered with their rapport and prompted apologies, remorse, anxiety, and guilt. Hick's volatile personality, given to highs and lows, and her unrestrained outbursts when she was overworked or irritated contrasted with the controlled dignity that was always characteristic of Eleanor. As Eleanor came to terms with her position as first lady and became an admired public figure in her own right, Hick's own career plummeted. No longer a nationally respected reporter, she never found a comparable place in the maelstrom of Washington politics.

A turning point in their relationship occurred during the summer of 1934, when they took a trip to Yosemite National Park. Instead of the private, woodsy retreat they had envisioned, their stay at the famed Ahwahnee Hotel inside the park turned into a nightmare of reporters, tourists, and forest rangers—all of whom wanted to photograph Eleanor Roosevelt. Moreover, the altitude didn't agree with Hick, who was suffering from diabetes and the negative effects of smoking. Though a decade younger than Eleanor, she could not keep up with her vigorous friend, who was hardened by daily horseback riding and a lifetime of physical discipline. When Hick verbally erupted at a gaggle of tourists, Eleanor was forced to pull her away from the crowd and calm her down in private.

It was an incident that Hick would remember, ashamed, in the biography titled *Eleanor Roosevelt, Reluctant First Lady* that she wrote in 1962. Significantly, the book ends with the last day of their 1934 Yosemite vacation.[247] Their friendship would continue for almost three more decades, but never with the passion of the first two years.

Hick and Eleanor's correspondence reveals the double portrait of two women who loved each other when it was not safe for women (or men) to go public with their same-sex feelings. It is not surprising that Hick censored some of their letters and may have destroyed others entirely before she turned them over to the Franklin D. Roosevelt Presidential Library. Eleanor and Hick knew that their mutual love would not have been acceptable to their contemporaries.

Eventually the intensity of Eleanor's feelings lessened, and Hick found herself in a painful position. Hurt feelings, canceled appointments, apologies, Eleanor's offer to return Hick's ring—all signaled the deterioration of their relationship. Nonetheless, during Franklin's second presidential term they established a precarious equilibrium that would continue for years.

The muted affection between them never disappeared, although they expressed it in different ways. When Hick found herself plagued with health and financial troubles, Eleanor came to her aid with checks and gifts. When Hick's beloved dog Prinz died in his fifteenth year, Eleanor sent her an English setter puppy. Hick continued to encourage Eleanor in her humanitarian activities, as well as in her public writing, which she sometimes edited. In the early 1940s, each acquired a new friend: Hick formed a special friendship with Marion Harron, a US Tax Court judge ten years younger, and Eleanor became close to Joseph P. Lash, a liberal intellectual twenty-five years her junior.

## Joe

Eleanor first met Joe Lash in the fall of 1939, when, as a left-wing student leader, he was called to testify before the House Un-American Activities Committee. Eleanor sat through the entire proceedings as testimony to her support for young activists, and after they were over, she invited Lash and five of his friends to the White House for dinner. The following summer he spent an entire week at Val-Kill, which grounded their lasting relationship—"as close a relationship as I ever knew Mother to have," in the words of Eleanor's daughter, Anna.[248]

Lash's later analysis of their friendship tells us a great deal about Eleanor at this stage of her life. With her children grown up and her husband immersed in public affairs, she had "a deep unquenchable longing to feel needed and useful . . . to have people who were close, who in a sense were hers and upon whom she could lavish help, attention, tenderness. Without such friends, she feared she would dry up and die."[249] Joe was more than willing to play the role of the needy younger friend.

Throughout World War II, Eleanor and Joe corresponded, and while he was stationed in the Pacific she kept a motherly eye on Joe's future wife, Trude. After the war, Eleanor and Joe cofounded Americans for Democratic Action, a liberal anti-communist organization.

At the time of Franklin's sudden death in April 1945, Eleanor could count on her friends for comfort and companionship. Isabella always remained in touch, even if she and Eleanor rarely saw each other, and Tommy continued to work for Eleanor until Tommy—like Isabella—died in 1953. Joe Lash and Eleanor stayed closed until she passed away, whereupon he became the first major chronicler of her life.[250] Hick and Eleanor saw each other and corresponded regularly until Hick received a telegram in 1962 announcing the death of Mrs. Franklin D. Roosevelt and inviting her to the church service in Hyde Park.

## A Widening Circle

Even before Franklin's death, Eleanor had been extending her friendship circle to include people outside the WASP milieu of her youth. Some of these new friends, received grudgingly by Nan and Marion at Val-Kill when they were still part owners, were working-class reformers, labor organizers, and people of color. Her relationship with Pauli Murray is a case in point.

In 1939, after President Roosevelt received an honorary doctor of law degree from the University of North Carolina, Murray felt compelled to write him a letter, with a copy sent to Eleanor. She pointed out that in his acceptance speech, which praised the university for its "liberal thought," Roosevelt had overlooked the fact that "Negroes" were still not admitted there. She herself had just been turned down by its graduate school. To Murray's surprise, she received a reply signed by Eleanor Roosevelt, encouraging her in her aspirations though advising her to "fight hard with concilia-tory methods." The letter inaugurated a relationship that was to last until Eleanor's death.[251]

Having been refused admission to white institutions, Murray went to Howard, a historically black university, to study law. In April 1944, during World War II, as she was preparing to graduate from law school, Murray and several other students slipped into the white-only Thompson's caf-eteria near the Tidal Basin in Washington, DC. When the employees refused to serve her, she sat down anyway with her empty tray, at a table with the other black students who had been refused service. They remained there silently, without responding to taunts or jeers. Outside, another group of stu-dents walked a picket line, carrying a placard that read: "We Die Together. Why Can't We Eat Together?"

Murray wrote to Eleanor to describe the protest, know-ing she would approve of their nonviolent tactics following the methods of Mahatma Gandhi, which were taken up by the Civil Rights Movement. Eleanor was always a temper-ing presence in the life of her firebrand friend. Still, Murray would write—after Eleanor's rousing Madison Square Garden appearance in May 1956, with Martin Luther King, Jr., Congressman Adam Clayton Powell, Jr., and Autherine Lucy, the first African American admitted to the University of Alabama—that when she was "riled up," Eleanor was "a bit of a firebrand" herself.[252]

In 1961, when Eleanor was asked by President Kennedy to

chair the Presidential Commission on the Status of Women, she arranged for Murray, by then a noted civil rights attorney, to draft the Commission's study. As her own health deteriorated and her energy flagged, Eleanor passed the torch to her younger friend, fully trusting in Murray's ability to keep the flame burning.

Special friends propelled the arc of Eleanor's adult life. They gave her the nurturance and mentoring that enabled her to remake herself as an entrepreneur, a teacher, and a social activist. She became a sought-after public speaker and the author of well-paid publications. She developed into a flourishing political figure, an ally of the downtrodden, a spokesperson for women, a voice for interracial harmony, and a champion of world peace. Her friends believed in her and emboldened her, and she in turn emboldened them.

Eleanor Roosevelt's influence was such that in 1946 President Harry Truman appointed her a member of the American delegation to the United Nations General Assembly. From that platform she led the crafting of the first Universal Declaration of Human Rights and went on to become, in the words of Harry Truman, "the first lady of the world."

Her example is too grand to be considered representative, and yet it does say a great deal about the opportunities that presented themselves to American girls and women during the first half of the twentieth century. Most girls living above the poverty level were expected to attend high school, and a certain percentage of them went on to college. Women also began to enter the workforce in growing numbers at this time, especially during World War II. Some women even found their way into politics and, like Frances Perkins, into positions of power. It is unlikely that women would have achieved such prominence during Franklin's presidency without Eleanor's prodding. However exalted, Eleanor's story offers a microcosm of women's multifaceted progress during seven crucial decades of American history. More than any other woman of her era, Eleanor Roosevelt made the most of her situation as a political wife. Empowered by friends, then mentoring others, she deserves to be remembered as a friend of humankind.

# ELEVEN

# FROM COUPLEHOOD
# TO SISTERHOOD

*men come*
*and go. your friends*
*stay. women*
*stay. mom*
*said.*

—ALMA LUZ VILLANUEVA, *MOTHER MAY I?*, 1978

*Charity and Sally are stitched together with a thousand threads of feeling and shared experience. Each is for the other that one unfailingly understanding and sympathetic fellow-creature that everybody wishes for and many never find.*

—WALLACE STEGNER, *CROSSING TO SAFETY*, 1987

*Before there were therapists, there were girlfriends.*

—CHRISENA COLEMAN, *JUST BETWEEN GIRLFRIENDS: AFRICAN-AMERICAN WOMEN CELEBRATE FRIENDSHIP*, 1998

T HE MARRIED COUPLE HAS BEEN the social bedrock throughout American history. When mass media began to shape the way people viewed their lives, the American public was bombarded with images of women and men in blissful or sparring pairs. Radio shows such as *The Romance of Helen Trent* and *Our Gal Sunday* serialized the adventures of women seeking permanent romantic attachments. Screwball comedies such as *It Happened One Night* (1934) and *His Girl Friday* (1940) depicted the battle of the sexes, and the blockbuster novel *Gone with the Wind* (1936) offered women the fantasy of marrying either the angelic Ashley or the rakish Rhett Butler. Only Grant Wood's satirical depiction of the American couple in his famous painting *American Gothic* (1930) seemed to suggest that there was something grimly funny about compulsory couplehood.[253]

Female friends played a minor role in the media, and when they did, they were often portrayed as rivalrous and treacherous. *The Women* (1939) is a classic example of Hollywood's presentation of women as wealthy bitches, ruthless in their efforts to steal husbands, lovers, money, and social positions from other women. *All About Eve* (1950) is the story of an aging star betrayed by a younger actress who eventually undermines her mentor's career and personal life. There has historically been little in popular culture to suggest that many women have depended on one another for emotional and material support in the face of unhappy marriages.

Once they had "nabbed" a husband, women were expected to be satisfied with their lot—especially during the

1930s, when the specter of the Depression hung over everyone. God forbid that a married woman should continue working and take a job away from a man! In a 1936 poll that asked whether a married woman should keep a full-time job, only 35 percent of the respondents said yes.[254] Despite the gains that working women had made during World War II, when their services were direly needed, pundits told them both before and after the war that homemaking was to be their full-time job. Many women balked at these restrictions and continued working outside the home, but many more tried to find fulfillment as housewives, mothers, and community members.

## Couples and Couples

In the 1930s and 1940s United States, friendship between two couples became popular. Wives were expected to be able to hold up their end of conversation and games in the foursome, and to serve refreshments for the visitors. Middle-class couples regularly played bridge with other couples, swapped dinner invitations, or met during the summer at the same vacation spot. Author Diane Johnson in her delightful memoir, *Flyover Lives*, recalls how her father, a high school principal, and her mother, a full-time housewife, had a core group of couples in their Moline, Illinois, neighborhood with whom they played bridge, poker, gin, and golf: "The Bosses, Martins, Gills, and Lains [her parents] made two tables for bridge, though they sometimes played poker instead—and always for money, small stakes . . . They all drank quite a bit, highballs or martinis, and guests sometimes had to be seen home, but this was only down the street, and on foot."

With her women friends Johnson's mother had two bridge clubs of her own, scheduled in the afternoon since none of them worked. She had been an art teacher when she came to Moline, but she "had been obliged in the custom of the day to stop teaching when she got married."[255] From the child's perspective, the friends were all fond of one another, even if they sometimes joked about who drank too much or who played bridge erratically. Whatever secrets were hidden behind this picture of small-town harmony were unknown to the future writer.

Author Wallace Stegner chose the theme of the friendship between two couples for his semiautobiographical novel *Crossing to Safety* (1987). In the novel, Stegner traces the interconnections of two college professors and their wives over a period of forty years, from the 1930s to the 1970s. *Crossing to Safety* represents a remarkable paean to that special form of friendship that emerges when four people all care for one another, as individuals and as couples.

Charity and Sally are stitched together with a thousand threads of feeling and shared experience. Each is for the other that one unfailingly understanding and sympathetic fellow-creature that everybody wishes for and many never find. Sid and I are close, but they are closer . . . What I am sure of is that friendship—not love, friendship—is as possible between women as between men, and that in either case it is often stronger for not having to cross lines. Sexuality and mistrust often go together, and both are incompatible with *amicitia*.[256]

Almost enviously, Stegner lauds the relationship between the two women for its seamless intimacy. We have come a long way from past male writers who considered women incapable of the loyalty attributed to men. And it is interesting to note that Stegner decisively separates friendship from sexual love, under the assumption that the latter will always be inimical to the former. This question has always dogged the discourse on friendship, whether it be that of ancient Greece or contemporary America.

In the 1950s, friendship between couples formed the grid for two of the most successful television sitcoms of all time: *I Love Lucy* and *The Honeymooners*. In the first of these two legendary programs, the members of the primary couple, Lucy and Ricky Ricardo, were played by Lucille Ball and Desi Arnaz (who were offscreen spouses). Fred and Ethel Mertz, their landlords and best friends, lived in the same building on the Upper East Side of Manhattan. They all displayed the gender stereotypes of their era. Thus Lucy, a daffy housewife and devoted mother, is ridiculed for wanting to work outside the home and enter show business like her bandleader husband, Ricky. Though she exhibits few of the requisite talents, she is supported against Ricky's and Fred's constant put-downs by Ethel, a former vaudeville star. The cast's skillful interplay of two couples locked in a zany friendship made for award-winning entertainment.

*The Honeymooners*, which also aired for the first time in 1951, featured two working-class couples. Living in a modest Brooklyn apartment, Ralph Kramden, a bus driver portrayed by Jackie Gleason, and his wife, Alice, played off their neighbors Ed and Trixie. *The Honeymooners* demonstrated that four-way friendships were not restricted to the Upper East Side and that they could find their way into the lives of modest folk. In both cases, the foibles of friendship between couples lent themselves to comic hilarity.

## 1950s

However idiosyncratic, sitcoms, movies, and novels reflected the mentalities of their era. In the fifties, when Americans were returning to postwar "normalcy," women and men married at a younger age than at any other time in the twentieth century. The median age at first marriage for women dropped precipitously between 1940 and 1950, from 23 to 20.5, and for men from slightly over 25 to 24.[257]

It was not unusual for teenagers to seriously date and then marry their high school sweethearts. Without the social sanction of premarital sex available today, young people rushed into marriage as soon as the man had a job in sight. A Wellesley graduate from the class of 1954 remembers that it was de rigueur to get engaged by one's senior year of college, marry upon graduation, and work until the first baby arrived, usually a year or two after the wedding. Afterward, women were expected to find satisfaction in family life, homemaking, and, in the words of poet Maxine Kumin, "weekends of enforced exchanges with other couples."[258] Friends were found among one's neighbors, with common interests in children and the PTA. A married woman determined to pursue further studies or a career was seen as "ambitious" and probably devoid of the feminine traits that marked true women. Welcome to the 1950s version of Victorian womanhood!

The rise of American suburbs offered housewives opportunities to form new female enclaves. It was easy to make friends with women of approximately the same age, income, and type of house. Many formed friendships through their children's activities and continued those friendships even as the children grew up, married, and produced children of their own.

1950s suburban communities tended toward all-white, middle-class homogeneity, with nonwhite and lower-income families excluded from the scene. Working-class families had their own patterns of friendship, which Mirra Komarovsky studied in 1958 and 1959.[259] She noted the rigid separation that existed between the husband's work life and his home life. The great majority of wives had no contact at all with their husband's job mates; whatever friendships the husbands made at work did not extend to their wives. This situation contrasted with that of middle-class "corporation wives," whose social lives often revolved around their husbands' careers. But many couples in Komarovsky's sample did socialize with other couples, 58 percent on a weekly or monthly basis. Such friends, some dating back to elementary or high school, would arrange to spend evenings together playing cards or watching TV. Sometimes they went out bowling, swimming, or hiking.

Only 17 percent never had visits with another couple. These statistics tell us that the ideal of couple friendships, visualized in *I Love Lucy* and *The Honeymooners,* did indeed make its way into working-class families, though to a lesser degree than in higher socioeconomic groups.

Blue-collar wives found friends among other women with similar concerns. They might form social clubs with six to ten members, meeting weekly at one another's homes to play cards or sew. Occasionally they put aside enough money from their dues to have a night out for dinner and a movie. Working-class women were active members of Mr.-and-Mrs. church clubs, the PTA, bowling leagues, and singing societies. On Komarovsky's list of ten activities rated according to enjoyment, "having friends over" rated third, after watching TV and playing with one's children.

In the early years of marriage, blue-collar husbands and their wives named their spouse as their favorite companion, but after seven or more years, the older wife named friends and relatives with increasing frequency in her list of preferences, whereas the older husband curtailed his extrafamily associations. There does seem to be a tendency for men of all classes, as they grow older, to depend increasingly on their wives for friendship, even as their wives find more friends outside the family circle.

*Friendship and Feminism*

In 1961, President John F. Kennedy appointed a Commission on the Status of Women, chaired by Eleanor Roosevelt. In

1963 (a year after Eleanor's death), the commission produced a report documenting discriminatory practices against women in government, education, and employment, and recommending means to correct them. Following suit, individual governors appointed their own commissions on women. Like New Women and early feminists, women were once again talking to one another—and to men—about the obstacles that blocked them from full participation in American society.

That same year, Betty Friedan published *The Feminine Mystique*, a bestselling account of the conflicts middle-class housewives were experiencing. Friedan's intimate style endeared her to a generation of white suburbanites, who were beginning to suspect there was more to life than the latest Westinghouse oven. Friedan would go on to become a founding member of the National Organization for Women and a leading force in the women's liberation movement.

In 1968, radical feminists picketed the Miss America contest. The protest was directed against the pageant because it epitomized the artificial standards of beauty imposed on all women. Demonstrators threw false eyelashes, high-heeled shoes, and girdles into trash cans, but, contrary to press reports, they did not burn brassieres. Nonetheless, the term "bra burners" stuck, and the women's liberation movement came to national attention through a largely negative press. Carol Hanisch, one of the organizers and demonstrators, described how the idea had originated at a "consciousness-raising" session: "We decided to go around the room with each woman telling how she felt about the pageant . . . From our communal thinking came the concrete plans for the action. The original planning group agreed that the main point in the demonstration would be that all women are hurt by beauty competitions—Miss America as well as ourselves."[260]

Consciousness-raising groups, like the one that led to the Miss America protest, were dedicated to the proposition that the personal is political—that one's personal problems are related to the societal structures that surround us. No, Julia, it is not because you are weak that your husband pushes you around. He pushes you around because society tells us that males are superior to females and that men should be in control. No, Patricia, you are not a freak for wanting to make furniture. Jobs should not be assigned on the basis of biological sex but chosen according to one's individual talents and inclinations. No, Margaret, you are not bizarre because you don't want to have children. We don't have to accept the outdated belief that mothers are the only "true" women. Consciousness-raising groups gave birth to a new form of friendship labeled *sisterhood*.

Carol Hanisch was already using the term *sisterhood* in her analysis of the Miss America pageant. She worried that the protest may have "harmed the cause of sisterhood" because it made beautiful women come across "as our enemy instead of as our sisters who suffer with us."[261] She wanted future actions to reach as many women as possible, with a clear message of all-inclusive sisterhood.

In 1970, *Sisterhood Is Powerful*, an anthology of feminist writing edited by poet Robin Morgan, appeared on the scene. It quickly became a bestseller despite attacks from reviewers, who reviled the strident, shrewish, and malcontent tone of its authors. It also inspired many women to become involved with the women's movement and to change their own lives.

*Sisterhood* became the catchword for female friendship, implying that all girls and women, even those unrelated by blood or marriage, should treat one another with the affection and loyalty expected of siblings. The word took on a bold political ring in the newly coined phrase "sisterhood is powerful." United as sisters, women could collectively achieve societal change that would benefit each person individually. Conversely, what one did on a personal level had political implications. When a wife involved her husband in the care of their children or in cooking dinner, the personal would reach beyond just one household into the larger society.

With their slogans, protest marches, and consciousness-raising sessions, second-wave feminists called attention to the specific needs of girls and women, and treated many aspects of their lives—including their friendships—with new respect. It was no longer politically correct to cancel an appointment with a girlfriend because a boy rang up and asked for a date. A woman could ask her husband to stay home with the children while she went with a friend to hear Sweet Honey in the Rock. The stock in female friendship was definitely on the rise.

Feminist theologians Carol Christ and Judith Plaskow went so far as to call the women's movement a "religious experience." They recast consciousness-raising as an embryonic ritual that strengthened its members and endowed them with a communal sense of mission to spread "the good word" to others. The traditional Judeo-Christian goals of illumination, self-transformation, and salvation could be actively pursued within a supportive community of female friends: "Calling each other 'sister,' feeling a new freedom to touch and hug one another are concrete expressions of the new bonds between us."[262]

Of course, not all American women wanted to be sisters, and many resisted the idea that female friendships should be on a par with kinship. Phyllis Schlafly led the antifeminist campaign and devoted her energies to opposing its aims. In particular, she lobbied against passage of the Equal Rights Amendment because she valued the fundamental differences between men and women and believed that full equality between them would not benefit either. Schlafly did not agree with the basic tenet of the women's liberation movement: that a woman's personal predicament is related to society and that it can only be solved collectively. To Schlafly, feminist sisters were faceless confrontationists, imprisoned rather than liberated by their negative view of the world. Schlafly offered a different view, that of the "positive woman," and with her organization, the Eagle Forum, she gave a voice to a more conservative population who were satisfied with familiar patterns and unconvinced that women were oppressed.

In spite of such backlash, feminist ideas filtered into society at every level. A Harris Poll of December 1975 reported that 63 percent of the women interviewed favored changes designed to improve the status of women. The list of organizations that supported ratification of the Equal Rights Amendments included not only feminist groups like NOW but also mainstream organizations such as the National Federation of Business and Professional Women's Clubs, the American Association of University Women, the National Council of Negro Women, and the YWCA. All those women's clubs, founded in the late nineteenth or early twentieth century, were still in full force and still pushing for women's rights.

In 1977, the International Women's Year conference took place in Houston, Texas, before two thousand delegates and twenty thousand guests. Three first ladies—Lady Bird Johnson, Betty Ford, and Rosalynn Carter—endorsed the ERA. They held a lighted torch that runners had carried from Seneca Falls, where the first women's rights conference had been held in 1848. The relay runners included Susan B. Anthony II, tennis star Billie Jean King, congresswoman Bella Abzug, and Betty Friedan. Delegates stood up or remained seated to indicate their vote. The only time that the entire delegation stood up was to approve the motion that all women, married or not, should be allowed to have bank accounts and credit in their own names. Economic considerations united every single woman.

### Comradeship Among Divorced Women

Carolyn See, born in 1934, wrote a remarkable account of a friendship that started in junior high school and lasted for half a century. During that time, she and her friend Jackie Joseph—like so many other girls and women—were marked by the upheavals of second-wave feminism. As girls, both Carolyn and Jackie lived with their divorced mothers in a lower-middle-class Southern California neighborhood. As See tells it:

> Here is what Jackie and I had in common: we were alone in the world, living with mothers who, on a good day, didn't like us very much. On bad days, they hated us with all their strength . . . What Jackie and I had in common was that we were poor, so *poor*! My mother worked as a typist. My dad, when he sent the child support, usually had his girlfriends pay it, out of their waitressing wages . . . Jackie's mother, for a while, actually owned a liquor store on Skid Row. She'd drive us—fourteen-year-olds—downtown to work for her.

The girls bonded over the shared misery of poor daughters with divorced mothers unwilling or unable to provide the protection that children crave. Nevertheless, they were good students in school and made friends with two sisters, Joan and Nancy, whose wealthy Jewish family offered them previously unknown luxuries: Carolyn and Jackie borrowed much of their high school wardrobes from Joan and Nancy. The four of them continued their friendship on and off for much of their lives.

By the time they were twenty-five, all four of the friends had married—this was the late fifties, when women married young and didn't live with men except as husband and wife.

By the early seventies, three of the four women were divorced, Carolyn twice. Sharing their divorced lives, Carolyn and Joan bonded through laughter as well as tears:

> Joan and I lived for some years in the harrowing land of divorce. (What women don't remember to say very often is that, as you get used to it, it's a fair amount of fun.) Many evenings she and her three kids and I and my two kids and whatever semi-boyfriends we could round up would drive on out to a great restaurant called the China Palace. The kids would go hog wild and we'd giggle and laugh and fall about, and nobody would get on our case. We could play! Nobody could order us around!

Jackie, too, parted from her husband, when he took up with a younger woman. She had been so busy raising their children, acting and singing, and taking care of her dying mom that she had scarcely noticed his earlier affairs. After her divorce, she decided she was going to change the way people thought about women living alone with children, so she founded a group called LADIES (Life After Divorce Is Eventually Sane).

Carolyn grimly recalled their situations: "Jackie, Joan, and I ended up like many divorced women—without an income, with children who yearn for their fathers with a terrible yearning and are ashamed to be seen in (just) your company." Welcome to American society of the 1970s and early 1980s, when divorce rates hovered around 50 percent.

Divorce often had the negative consequence of leaving women with less income than their ex-husbands, and with the primary care of their children. It became clear to many women—like Carolyn, Jackie, and Joan—that living on what they earned and what their husbands were supposed to pay in child support would barely keep a roof over their heads.

Yet an unexpected positive consequence of divorce could be found in the sense of comradeship that sprang up among divorced women. With nobody to boss them around, Carolyn and Joan were free to enjoy themselves in new, fun-filled ways. Fun was not something they had associated with their marriages. And Jackie found collective support in LADIES, putting them on television and presenting a model of solidarity among divorced women: "Listen!" she said. "It's bad for all of us. But we can get over the personal hurt, and we can change the system." The personal had indeed become political.

Carolyn went on to marry for a third time, and this time she got it right. By 1992 she and her "wonderful man" had been together for nineteen years, and her friendship with Jackie had lasted for forty-six. As Carolyn looked back, her choice of words bore witness to the tumultuous years of women's liberation. *Patriarchy*—that feminist bugaboo—had found its way into her vocabulary when she wrote: "This is a story of female friendship in a patriarchy, except that when Jackie and I were little, we couldn't even afford a patriarch!"

Carolyn See summed up her lifelong friendship with a moving appraisal: "I count it a miracle that two desperately poor, desperately lonely semiorphans could, day after dopey day, put together a friendship that can last up to (and further than) half a century. I know the world is better because Jackie Joseph lives in it. I'm lucky, so lucky, to know her."[263]

The feminist movement consciously emphasized female friendship as never before in American history. Whereas in the past, a woman's friends had been expected to play supporting roles in relation to the marital couple and the family, now they were placed squarely on the map of human relations. The importance of friends was recognized for both personal happiness and the well-being of society. Feminism encouraged women to value one another in a new way and to work collectively for equal status with men.

For this they would need the involvement of many women willing to treat one another as sisters, and that population could not be limited to the white middle class living in New York City. Sisters had to be found elsewhere in America, among black women, Asian Americans, Latinas, and other women of color. They had to believe, like Carolyn, that women as friends could make the world a better place for everyone, regardless of race, religion, ethnic background, sexual orientation, or socioeconomic status. And for a time, in the 1970s and 1980s, it looked as if that vision could be realized, especially for white, middle-class women.

## Black Sisters

The issue of race within the women's movement was vexed from the start. African Americans and other women of color did not initially see themselves as sisters with white women, and vice versa. They had reason to distrust the idea of friendship with members of a race that had brought their ancestors to America as slaves and treated their descendants as inferiors. Strong female networks had always existed within African American communities, even though they were subject to domination by black men and the white world. During the turbulent Civil Rights movement of the 1960s, the first loyalty of black women was to their own race—to the men and women who marched in the streets, sat defiantly in "white-only" seats in drugstores and on buses, and endured the humiliation of being jeered at and spat upon by angry white spectators. It is true that many white men and women joined African Americans in their struggle, and that interracial relations have, on the whole, improved since then. But back in the 1960s, when the Civil Rights Movement was in full swing, the concerns articulated by middle-class white feminists were not echoed by most black women.

It took their own experiences for African American women to realize that patriarchy was as detrimental to them as it was to white women. Treated as subordinates by the black men in charge of political action, the women slowly became more receptive to feminist ideas. Perhaps the term *sister*, so prevalent in African American parlance, could be extended into *sisterhood* on a broader scale. Perhaps it was time for concerns about gender to vie with concerns about race.

Writing by black women in the 1970s and 1980s clearly reveals the presence of feminist thinking. The popular poet Nikki Giovanni, in her 1971 autobiography, *Gemini*, addressed

the subject of male domination head-on, using a style that reflected African American vernacular:

> And sometimes you say, that's all right; if he takes advan-
> tage of me, so what? At nineteen that's cool. Or maybe
> at twenty-three. But around twenty-five or thirty you say,
> maybe men and women aren't meant to live with each oth-
> er. Maybe they have a different sort of thing going where
> they come together during mating season . . . but living
> together there are too many games to be gotten through.
> And the intimacies still seem to be left to his best friend
> and yours. I mean, the incidence is too high to be ignored.
> The guy and girl are inseparable until they get married;
> then he's out with his friends and she's out with hers or
> home alone . . . She's just not the other half of him.[264]

Giovanni's conclusion, that women must count on their
female friends more than on their lovers or husbands, found
echoes in an entire generation of black women writers, including Toni Morrison, Alice Walker, Gloria Naylor, and Terry
McMillan.

In Toni Morrison's 1973 novel, *Sula*, the friendship between two women serves as the central theme joining the
heroine's troubled adolescence in a Southern town and her
return home after years of wandering through America's big
cities. Toughened up by diverse relationships with men, Sula
came to realize that "she had been looking for a friend, and
. . . that a lover was not a comrade and could never be—for

a woman."[265] In this book, as in her others, Morrison never offers comforting solutions. Her women often end up in madness or death. Even friendship cannot always withstand demeaning circumstances.

In a 1983 interview, Morrison reflected on the novelty of the friendship motif in *Sula*:

> Friendship between women is special, different, and has never been depicted as the major focus of a novel before *Sula*. Nobody ever talked about friendship between women unless it was homosexual, and there is no homosexuality in *Sula*. Relationships between women were always written about as though they were subordinate to some other roles they're playing. This is not true of men. It seemed to me that black women have friends in the old-fashioned sense of the word; perhaps this isn't true just for black people, but it seemed so to me. I was halfway through the book before I realized that friendship in literary terms is a rather contemporary idea.[266]

Increasingly, after *Sula*, female friendship began to act as a prism through which black women writers constructed their stories. Gloria Naylor's 1982 novel, *The Women of Brewster Place*, offered a collective portrait of women living side by side on a lower-class, dead-end street. The relationships between mothers, daughters, and friends are played out against a shifting scene of fathers, lovers, husbands, and sons, who are rarely in the picture when the women need them—with the exception of Ben, the aging wino janitor. Naylor's world contains horrendous cruelty but also moments of sublime grace. When the young mother Ciel sinks into deathlike apathy after the accidental death of her baby girl, which occurs during a quarrel with the baby's no-good father, Ciel's older friend Mattie brings her back to life through determined acts of love. Mattie rocks Ciel and bathes her, as if she were a baby. She draws upon a fierce maternal love that will not let her friend die.[267]

The rocking and bathing are enacted with exquisite tenderness. Women normally do such things for their babies, and sometimes for the elderly, but taking care of a friend in this fashion is less common. Yet here it seems natural, necessary, an extension of the nurturance traditionally associated with women. This is not to say that men are incapable of such nurturance, or, conversely, that all women are. Many men take care of the physical needs of their babies, wives, lovers, and parents. But let's face it: such care normally falls to women. Even under less dire circumstances, girls and women are known to wash each other's hair, paint each other's nails, and massage each other's backs. There is often a physicality in the friendship between women that does not have to be sexual and that may, as in the case of Mattie and Ciel, have a healing force.

Alice Walker also published her Pulitzer Prize–winning novel, *The Color Purple*, in 1982. It, too, is based on the theme of women's friendships struggling to assert themselves in a world dominated by men. In the words of Sophia, one of the novel's main characters: "All my life I had to fight. I had to fight my daddy. I had to fight my brothers. I had to fight my cousins and my uncles. A girl child ain't safe in a family of men."[268]

Yet one woman character, Shug, manages to escape and live a relatively independent life as a blues singer. She forms a crucial friendship with the central heroine, Celie, who is the opposite of Shug and Sophia in every way. Celie is meek and submissive, first to her father, then to her husband, both of whom exploit her labor and her sexuality. Celie and Shug become friends when Shug is sick and Celie nurses her back to health. The nurse–patient bond between them develops into a sisterly-cum-erotic attachment, with Shug initiating Celie into the secrets of the female body. At one point, Shug persuades Celie to look at her own genitals with a mirror. This episode recalls the late 1960s, when *Our Bodies, Ourselves*, a revolutionary feminist text, encouraged women to examine their own bodies, with mirrors where necessary.

The transformative friendship between Celie and Shug is one of several female relationships that form the spine of *The Color Purple*. Another is Celie's connection to her sister, Nettie, who is sent away to Africa with missionaries because she refused the sexual advances of Celie's husband, Albert. Albert's cruelty continues long after Nettie leaves: he confiscates all the letters she sends to her poor, demoralized sister, Celie. When these letters come to light through the wiles of Shug and Celie, they help give Celie the strength to stand up to Albert and leave him. Yet for all the resistance against male oppression that energizes the novel, it is far from an anti-male tract. Walker has created a mythic saga in which black members of both genders find their common humanity through women's conduct as friends.

*The Color Purple* was made into a movie in 1985. Directed by Steven Spielberg, starring Whoopi Goldberg, and including such actors as Danny Glover and Oprah Winfrey, it was a huge critical and box-office success. More than any other book or film of its time, *The Color Purple* established women's friendships as a subject that the media could no longer ignore. Female friendship was not a reality belonging to black women or to white women; it was a truth to be universally acknowledged in countless novels, films, and TV shows in the following decades.

### Female Friendship on Film

Look up "female friendship in film" on the Internet and you will find lists of the ten or twenty or fifty best movies based on the theme of girls' or women's friendships. Some of the most frequently cited are *The Color Purple* (1985), *Beaches* (1988), *Mystic Pizza* (1988), *Steel Magnolias* (1989), *Thelma and Louise* (1991), *Enchanted April* (1992), *The Joy Luck Club* (1993), *Waiting to Exhale* (1995), *Antonia's Line* (1995), *How to Make an American Quilt* (1995), *The First Wives Club* (1996), *Romy and Michele's High School Reunion* (1997), *Divine Secrets of the Ya-Ya Sisterhood* (2002), *The Sisterhood of the Traveling Pants* (2005), *Volver* (2006), *The Wedding Song* (2008), *Bridesmaids* (2011), *Snow Flower and the Secret Fan* (2011), and *Frances Ha* (2012). A slew of more recent movies indicates that the appetite for female friendship flicks has not abated. Many of these films had first appeared as successful novels, standing out from the mass fiction that is often dismissed as chick lit. Some were peopled by African

Americans and others by Asian Americans, and some were produced in countries far from the United States. All established women's friendship as a force to be reckoned with in the contemporary world.

These movies pass the Bechdel test, which has three criteria: (1) the film must have at least two women who (2) talk to each other about (3) something other than a man. The test was invented by Alison Bechdel in her comic strip *Dykes to Watch Out For* (1983-2008) and has become a shibboleth for anyone hoping to promote more realistic images of girls and women in the media. To this end, some films have focused on less attractive aspects of female friendship, especially during the teen years. *Mean Girls*, a 2004 comedy, depicted the ways cliques can hurt girls who are not "in," or popular. Based on Rosalind Wiseman's popular self-help book for parents of adolescents, *Queen Bees and Wannabes*, the topic hit a nerve. Some social scientists attribute girls' interpersonal cruelty to their "superior social intelligence," which lets them target other girls' weak spots. Most—but not all—women grow out of it.[269]

### *Girlfriends on TV: From* The Mary Tyler Moore Show *to* Broad City

In the 1950s, *I Love Lucy* presented girlfriends in the context of two married couples. Society was not yet ready for two single gals to flicker through their living rooms unescorted by proper husbands. It was not until the 1970s, with *The Mary Tyler Moore Show*, that television featured a woman whose life centered on her work and friends. Mary's entourage included her best friend, Rhoda, and a less likable, snobbish friend named Phyllis. Mary's close relationship with Rhoda earned such iconic status that they became a hilarious trope twenty years after the show's demise in another iconic depiction of female friendship, the 1997 film *Romy and Michele's High School Reunion.*

In 1981, *Cagney & Lacey* depicted two no-nonsense New York City detectives who propped each other up in their professional and personal lives. Importantly, one was a working mother and the other a single woman navigating various romances. Brought back from cancellation due to fan outcry, it ran for seven seasons.

Striking a different note, and one quite relevant to a future generation of aging baby boomers, *The Golden Girls* (1985–1992) presented four retired women sharing a house. These women were all single, and the comedy centered on the hilarious ways they made a family together.

*Seinfeld* broke the sitcom mold when it first aired in 1989. Famously "about nothing," it featured a group of early-thirtyish friends and their silly daily interactions. One of the lead characters, Elaine, held best-friend status with the main character, Jerry. The backstory was that Elaine was Jerry's former girlfriend, but the sexual chemistry necessary for heterosexual romance had eluded them. Elaine, played by Julia Louis-Dreyfus; Jerry Seinfeld, played by himself; and Jerry's two wacky male buddies mirrored the societal changes that have allowed men and women to interact with each other as bona fide friends.

By the 1990s, American TV was ripe for one of the most popular sitcoms of all time: *Friends*, which ran from 1994 to 2004. The seed story centered on three women and three men, all members of a friendship group that functioned as the equivalent of a family. As a marker in the history of women's friendships, *Friends* was significant for the even footing on which female and male characters were placed. Rather than the zany foibles of a woman sparring with a more dominant male figure, as in *I Love Lucy*, or the breaking-the-mold status of the single career girl Mary in *The Mary Tyler Moore Show*, *Friends* reflected the way young people were, and are now, engaging in group friendships.

*Sex and the City* (1998–2004) also featured a friendship group, but this one was all about women (and occasionally their gay friends) who unfailingly supported one another in their search for romance, happiness, and new shoes. *Grey's Anatomy* (2005–) provided a needed counterpoint in its depiction of the serious best friendship between two highly accomplished medical women. As one critic put it, "When Meredith's husband feels a third person in their bed, he knows it's Cristina."[270]

Online, girls and women have found countless models for friendship among women comedians, such as Amy Poehler and Tina Fey in their *Saturday Night Live* skits. Like other world-famous power couples, these two sharp-witted women are frequently featured in celebrity journals as a real-life best-friends pair.

Poehler went on to celebrate women's friendship in the TV series *Parks and Recreation*. The relationship between Poehler's character and her best friend, Ann, like that between Mary and Rhoda forty years earlier, is central to the ongoing plot. The dialogue crackles with memorable one-liners, such as Ann's declaration: "Never send a husband to do a best friend's job."

Meanwhile, YouTube videos and formally produced television series have allowed powerful new female talents to surface in public viewing spaces and produce creative, often hilarious depictions of female friendships that do not revolve around men. Two excellent recent examples were both created by a real-life duo of best friends, Lennon Parham and Jessica St. Clair. Their first foray, cancelled after six episodes, was *Best Friends Forever*, and the second, which began in 2014, is *Playing House*. The characters do things best friends do: one jumps on a plane to aid a friend in desperate need; one physically pulls the other out of a seven-hour bath; one uses the other's boob as a lip-sync microphone. Sure, some of it's weird, but it's also very funny.

*Broad City*, a show featuring modern-day women remaining friends through life's most trying moments, started getting media recognition after debuting on YouTube in 2009. The show is a comic exaggeration of the real-life intense friendship between Abbi Jacobson and Ilana Glazer, which exudes a "genuine camaraderie."[271] As the *New Yorker* described this "bra-mance," it is "an unpretentious portrait of a friendship between women in which they don't undermine each other or fret over how they look or define themselves by whom they're sleeping with. The love affair at the heart of the show is between Abbi and Ilana."[272]

### Sisterhood Rising

Wherever we look (and we *are* looking!), we find signs that American girls and women are counting on one another for companionship and support, perhaps more so than ever before. *Sisterhood* may no longer have the political ring it did in the 1960s and 1970s, but it has entered into women's consciousness as a positive force, especially now that lifelong heterosexual couplehood has become more dubious. Whether they are married, single, divorced, or widowed, many women find in their friends, individually and collectively, ears receptive to their emotional needs and tongues offering empathetic advice.

In her book *Just Between Girlfriends*, journalist Chrisena Coleman presents a series of vignettes that embody the importance of friendship in the lives of contemporary African American women. Her upbeat messages: "We can't pick our family, but thank God we can pick our own friends." "A good friendship is a lot like a fine wine—it gets better with age and time." "When all else fails, I call my best friend. She has a way of wiping my tears and calming my fears." "Men come and go, but best friends are forever."

Like American women of every ethnicity—white, black, Latina, Asian American, Native American, or other—Coleman has a special circle of friends who meet regularly on a semiformal basis, calling themselves Sisters of the Yam. They throw birthday parties for each of the five members, often in their homes but sometimes in out-of-the-way places such as Martha's Vineyard or a honkey-tonk dance hall.

Commenting on the value of their group, one of the members said: "We are sisters and friends. We support each other and truly enjoy our network. The Yams are always there to make our lives more pleasant."[273]

*Sisters and friends.* Tens of thousands of American women today would comfortably use those same words to describe the members of their friendship circles. Countless all-women groups meet weekly or monthly for tennis or bicycling, basketball or yoga. Female church choirs throughout the nation can be counted on for Sunday inspiration, while amateur and professional a cappella groups enliven private and community events. Women's garden groups have a long history in the United States, as do literary circles that unite women around a book they are all supposed to have read. Some college alumnae make a point of meeting every year at a designated locale, despite the distance involved.

The Dear Sisters' Literary Group, organized by a contingent of "thinking" black women, began coming together in 1995 for monthly meetings in one another's homes. The women ranged in age from twenty-seven to seventy-two and identified themselves as Christian and "womanist." N. Lynne Westfield, one of the group's members, defined the gatherings as "laugh-fests": "The gatherings are women laughing with each other rather than at each other. Women laugh about men, jobs, White people, neighbors, preachers, race, gender, pets, hairdressers and wardrobes."[274] Shared laughter releases pent-up anxiety and expresses the joy Westfield experiences with her African American sisters.

The Mothers' Study Club of Cambridge, Massachusetts, originally founded by a small group of Harvard professors' wives, celebrated its centenary in 2014. One member remarked that she had enormous admiration for the intelligence, sensitivity, and generosity of the older members—women who hadn't had the professional job opportunities that she had had yet made their mark in response to the challenges of war, illness, multiple moves, and the crises that confront us all. Intergenerational friendship is only one of the many forms that sisterhood takes.[275]

*Sisterhood* has edged itself into everyday speech, alongside such terms as *brotherhood*, *fellowship*, and *fraternity*, to suggest a community of feeling—if not between all human beings, as the masculine terms imply, then at least between all women. What once denoted only a relationship between blood sisters or, by extension, a society of women in a religious order may now designate all women who share the same experiences, interests, or concerns. Having lost its militant, anti-male stance, sisterhood may be more pervasive today than it was fifty years ago and, in its own way, more powerful.

# THREE

# Face-to-Face in the Twenty-First Century

## TWELVE

## FRIENDTIMACY

*"The social world is led by women."*

—SHERYL SANDBERG, CHIEF OPERATING OFFICER OF FACEBOOK, 2012

*"None of us use [email] exactly the same way we did ten years ago; in another ten years we might not use it at all."*

—MIRANDA JULY, ARTIST, 2013

LIKE IT OR NOT, SOCIAL media has fundamentally changed the ways in which nearly everybody conducts their friendships,[276] but more so for women than for men.[277] Social media is more important to women in part because it can accommodate the expressions of affection and self-revelation that often characterize female friendships. These empathetic expressions contrast with the norm for man-to-man friendships, which by and large can exist without the intimate confessions women so often make to one another.

The increasing scarcity of women's disposable time has helped spawn the mushrooming of social media. Even in dual-income households where the husband sincerely tries to shoulder a fair share of domestic burdens, the "second shift" of housekeeper/mother duties is still more often than not borne by the wife. Consequently, women in the twenty-first century have reincarnated themselves as quintessential multitaskers. Social media provides critical tools for women who manage the domestic front and the job front but who still wish to maintain important friendships. As Facebook honcho Sheryl Sandberg notes, women do the majority of the sharing on Facebook. Whereas men generally use social media for research and status boosting, "the social world is led by women," according to Sandberg.[278]

### "Reach Out and Touch Someone," Redux

While "efficient friendship" might sound oxymoronic, social media allows busy women to maintain friendships separated by distance or conflicting schedules. A few girlfriends might use social media to schedule an impromptu get-together that never would have happened in an age of telephone tag.

Nor is efficiency a bad thing in expressions of caring for women living far from their good friends. Dear ones can reach out to one another through a screen. When strong female friendships have outlasted life's changes despite time and distance, a quick text or a post as simple as "Julia ate oatmeal from her red bowl today" speaks worlds. Similarly, the success of the mobile apps Instagram and Snapchat demonstrate that instantaneously transmitting a picture to a friend can indeed be worth a thousand words.

One busy woman recently expressed the feelings of millions of posting, tweeting, and texting women around the world: "I message constantly with close friends throughout the day. . . . We share everything because we like acknowledgement and zero judgment. Acceptance and love. Would I trade in these 500 texts per week for a sit down with the three of them once a week? Nope, because it wouldn't happen. Not only is it geographically impossible, these real time exchanges are cathartic and they squash loneliness like a bug. . . ." [279]

### Anybody Out There?

Beyond sustaining long-term bonds, social media has opened limitless possibilities for finding and cultivating new friendships. If a woman is dissatisfied with the social opportunities in her immediate surroundings, a networking site could lead her to that group of sister techies, or knitters, or quilters, or amateur rocket builders she's been craving. These sister enthusiasts might be just down the street, or halfway around the world.

Loneliness affects men as well as women, of course, but historically, women tied to domestic obligations have had fewer means of connecting with people outside their immediate circle. Social media offers women unprecedented access to potential friends. Consider the website group Moms Who Need Wine. The word *need* in the context of alcohol is a sort of in-joke among the women who follow this group. Most of their posts and tweets tend not to talk about lofty sentiments of global import. Rather, they express the typical daily adventures and frustrations of caring mothers. But on the group's Facebook page and in their Twitter interactions, they don't risk social disapproval. Their sister Moms Who Need Wine get it, period. For women who seek to bond around activities besides mothering and wine, social media offers just about any combination of demographics, socio-economics, gender-identity, and common interest from which a friendship group can be formed. And in the unlikely event the sought-after group cannot be found, it is relatively simple to start one with a new Facebook group, or a Twitter or Instagram hashtag.

One such specialty web-based group is SheWrites.com for literary women. Kamy Wicoff and Deborah Siegel started She Writes as a virtual room, or salon, where women who cared deeply about one common thing—their writing—could get together and share relevant information. Within this "circle of knowledge and support"[280] , friendships often develop online. The gestalt of the site mimics what Wicoff and her sister *sa-lonnières* found when they were gathered together physically: "Our members are in an exchange, rather than one person talking at a group through a megaphone. That is what our site is about. It's a national extension of that room."

Wicoff noted that even in virtual salons, friendships are formed that extend well beyond the site. When she needed relief from the expanding time commitment required to run SheWrites, her online colleagues rallied as friends. "So I put out the call: 'I really need help—would anybody be willing to guest-curate the blog to give me time to devote to my own work . . .' I was hoping for a few responses. Within [a week], I had seventy volunteers!"[281]

As with face-to-face friendships, online relationships can center on matters small or great. While the occasional tribulations of motherhood or a shared focus on writing may not sound the deepest emotional chords, some online connections do develop over events that affect people profoundly and lastingly. A poignant example was the widespread e-grief over the March 2015 death of Lisa Adams. A suburban stay-at-home mom, she had built a following of more than 15,000 by blogging and tweeting about her life as she was dying from metastatic breast cancer. People who had never met Lisa in person and yet considered her a close friend cried over their smartphones at the news of her death. They texted, tweeted, and emailed each other, seeking what people have always sought from friends: mutual solace and support after the loss of a dear one. In sympathy, hands were squeezed and hugs were exchanged, all virtually, and yet somehow, all for real.

### Virtual? Reality?

A fundamental feature of social media is the slippery meaning of *reality* in *virtual reality*. Even the tweeter who offers up her thoughts raw and uncut is presenting a virtual version of herself to her Twitter audience. More obviously, deceptions online are often intentional. The picture you use for your online dating profile might be you twenty pounds lighter or five years ago, or it might not be you at all.

Performance artist Miranda July, in a 2013 online "installation," explored the idea of creating personae in emails. The project[282] involved having minor celebrities and lesser-known people submit real emails they had sent in the normal course of their lives. July then emailed batches of these once personal messages to people who signed up to receive them as part of her project.

On her website, July wrote about the art of online privacy:

> I'm always trying to get my friends to forward me emails they've sent to other people—to their mom, their boyfriend, their agent—the more mundane the better. How they comport themselves in email is so intimate, almost obscene—a glimpse of them from their own point of view . . . Privacy, the art of it, is evolving . . . none of us use[s] [email] exactly the same way we did ten years ago; in another ten years we might not use it at all.[283]

The ways in which we sculpt our privacy online—what we choose to reveal and what we keep hidden—as well as the ways we manipulate our personae clearly affect the quality of our Internet friendships. The "truth" becomes more elusive than ever. Although people might manipulate or lie to others in person, the more sensory stimuli we receive face-to-face, the more our gut feelings can help us discern what is real.

According to Joe Navarro, a former FBI counterintelligence special agent who has made a life's work of observing nonverbal ways in which people communicate:

Nonverbal behaviors comprise approximately 60 to 65 percent of all interpersonal communication . . . Nonverbal communication can . . . reveal a person's true thoughts, feelings, and intentions. For this reason, nonverbal behaviors are sometimes referred to as *tells* . . . body language is often more honest than an individual's verbal pronouncements.[284]

Today, texting, online chatting, and tweeting have removed from many interactions a rich overlay of sensory signals such as tone of voice, gestures, body language, touch, and smell. The loss of these important sensory cues hides certain truths that are much harder to disguise in real life. For women, social media is useful for augmenting friendships, but the deep empathy that they find by looking in a friend's face and offering a friendly hug are lost behind the screen.

Then, too, the socially uninformed, especially among younger digital natives, often misunderstand, or fail to apprehend, the differences between online friendships and real-life friendships.[285] The reality of social networking sites is that they provide platforms for online personae to interact with other online personae. Importantly, such relationships can be ended with a click of an "unfriend," "unfollow," or "block" button. Breaking up like this constitutes a morally lightweight action. Certainly it flies in the face of Cicero's advice that a friendship "should seem to fade away rather than to be stamped out." The respect that Cicero demanded we pay to a friendship, even one that has turned sour, did not anticipate the tenuous connection inherent in being a Facebook friend.

Every aspect of online communication offers upsides and downsides. In addition to removing sensory signals, the remoteness of online communication allows for shallowness, spitefulness, and hate to proliferate all too easily. Then again, the buffer of a screen can sometimes have the opposite effect. Messages of love and support that may be difficult to say in person get communicated quickly online. For better or worse, these new worlds of friendship can just be an emoji away.

Facebook proclaimed in 2013 that Internet access is a human right. Perhaps. But lofty proclamations alone do not convert connectivity into connectedness, a fact that is leading many women to take online social networking back into the real world. Women are going online to take friendships offline.

### Going Online to Get Offline

In spite of the amazing range of groups online, the fundamental human drive for creaturely contact with other people has not vanished into the ether. Connection in person remains a priority for many women. Consequently, sites for face-to-face friendships are growing in popularity.

Scott Heiferman, founder and CEO of Meetup.com, which he started in New York City in the aftermath of 9/11, said, "Our goal was to create a web platform that helps people form local communities around whatever is important to them, using the Internet to get off the Internet and meet up."[286] Meetup.com is the largest site dedicated to local platonic meetups; its membership numbers indicate that online interactions can never fully replace socializing in person.

Although Meetup.com itself is not gender-specific, many groups within it are formed exclusively by and for women in highly diverse combinations of ages, ethnicities, and interests. The difference from most website groups is that Meetup.com participants meet in-person and are in the same geographical area. There are meetup groups specifically for Latinas, African American women, Asian women, women engineers, women artists, old women, and young women. Importantly, women can find any mix of potential friends based on common interests, regardless of ethnicity, gender identity, or cultural background. For instance, some lesbians seeking friendships with like-minded women might find camaraderie with the "East Bay lesbian friendly activity group over 40." These gals are interested in group activities such as bike riding, hikes, small-dog walks, and dinner events. Sexual orientation defines relatively few Meetup groups. Other women seeking friends like themselves might try Sista's Book Social Club or Bay Area Divas of Color, among myriad other self-defining groups across the country.

Several entrepreneurs have gone a step beyond the common-interest focus of Meetup.com to design sites that highlight friendship first and foremost. There is a potentially gigantic target market for women seeking deep, nonsexual friendships with other women. Three of the biggest sites for this market are GirlFriendCircles.com, GirlfriendSocial.com, and SocialJane.com.

These girlfriend-specific sites have identified a big need among women today: even when two women are already friends, life can get in the way. Friends move away, kids grow up, married couples divorce, people die. For myriad reasons, women often wish for a new friend. And well-established social channels that exist for finding mates don't necessarily apply to women looking for platonic girlfriends. Buying a drink for another single woman at a bar sends a sexual message. Walking up to an unknown group of laughing women and trying to join in is regarded as awkward, if not downright pathetic.

A woman may have had close friends in a former phase of life, but now, in transition, she comes up empty. The question looms: how do friends find one another among today's fluid lifestyles?

Janis Kupferer, founder of SocialJane.com, remembered feeling lonely in a city where she'd moved for a new job. She browsed through the dating site Match.com and sometimes clicked on other women's profiles, wishing she could make friends with some of the women she read about. Why, she wondered, wasn't there a site for making platonic girlfriends?[287] And so she founded SocialJane.com, which allows women to be highly specific about the kind of potential friend they would like to meet. If you're looking for a Latina running buddy, or an African American professional who owns a dachshund, or a fellow reader of Nabokov novels in Russian, you'll probably find her on one of the girlfriend sites.

The thousands of women who have signed up for friendship sites want real contact with real women, in real time. At first glance, the sites function like online dating sites. Candidates submit their profiles, listing what they are looking for in a friend. Then, either they sift through profiles on their own and make contact if they detect a potential match, or the service matches them with similar women within a given geographic area. GirlFriendCircles organizes introductions in small groups, usually at a coffeehouse, so that candidates are presented with several options.

Girlfriend meetup sites have not swept the globe with exponential growth. Nevertheless, Shasta Nelson, founder of GirlFriendCircles, believes that the genre will mirror the trajectory of online dating sites:

> When online dating started, the idea got a lot of resistance. 'I'm not that desperate,' people said. With girlfriend sites, people at first say, 'What? I have to pay for friends?' But just like people used to be ashamed of online dating, now it's almost abnormal not to be on it if you'd like to date someone. Female friendship sites are going to follow that trajectory.[288]

A GirlFriendCircles member recounted on the site's blog her own success at finding friends. Shoshana based her decision to join the women's group after she had found her boyfriend through a matchmaking site:

> While on vacation in Mexico last October celebrating my thirtieth birthday, my boyfriend and I were discussing how hard it was for me to not have my friends living nearby. I grew up in Orange County but have been in Los Angeles

for several years—it's amazing what a little distance can do to friendships. Between graduate school and working full-time for many years, with no time to go out or to meet people/foster new friendships, I was feeling the absence. My boyfriend has nice friends who I enjoy spending time with, but I really felt like I should have my own friends as well. He jokingly suggested that there should be a site that exists to make friends, similar to a dating site (we met on JDate almost four years ago). So I pulled out my iPad and searched on Google. I found GirlFriendCircles.com, liked what I saw, and decided that I would join when I got back to LA.[289]

So social media may well drive more women seeking friends to use the Internet to get off the Internet. Social media will evolve until the "next big thing" replaces it. We can be sure that women will continue to reinvent their friendships as new technologies shift the cultural paradigm once more.

*THIRTEEN*

## GIVE AND TAKE: FRIENDSHIP
## IN A MARKET ECONOMY

*"I remember Kellie leaning across that desk and saying, 'Whatever you do, do not quit. You can do this.' . . . It's one of those pivotal moments . . . It meant the world to me."*

—KIRSTIN GROOS RICHMOND, COFOUNDER, REVOLUTION FOODS

*"We might say of friendships that they are a matter not of diversion or of return but of meaning . . . It is precisely this noneconomic character that is threatened in a society in which each of us is thrown upon his or her resources and offered only the bywords of ownership, shopping, competition, and growth."*

—TODD MAY, "FRIENDSHIP IN AN AGE OF ECONOMICS," *NEW YORK TIMES* ONLINE, 2010

FEMALE FRIENDS CARE FOR ONE another in an ongoing dance of give-and-take. Today, as more women find themselves with less free time,[290] the market economy fills some of the needs that friends and family used to answer. Take, for example, care for the sick and elderly, or living arrangements in shared housing, or interactions between a professional mentor and her protégée. There are undoubtedly economic motivations in these transactions, but there are also many features of female friendship that we have seen in a wide variety of times and cultural circumstances.

### Birds of a Feather

In one way or another, economics colors most aspects of our lives—where we live, whom we know, what we do. But can we really strip down all our human relationships to the so-called "dismal science"?[291] Thankfully, no. Empathy, loyalty, and affection—these remain priceless sentiments that bind people to one another. Nevertheless, economics is at work in our choice of friends. All else being equal, people tend to connect with those whom they regard as similar to themselves. Calling this phenomenon "homophily," social scientist Nicholas Christakis puts it this way: "Whether it's Hells Angels or Jehovah's Witnesses, drug addicts or teetotalers, Democrats or Republicans, stamp collectors or bungee jumpers, the truth is that we seek out those people who share our interests, histories, and dreams. Birds of a feather flock together."[292]

Very generally speaking, we tend to form friendships with people in a socioeconomic class similar to our own. This is largely because we find one another at a common time and place, and discover areas of life interest that overlap. And though we try to turn a blind eye to our friends' net worth, financial discrepancies can cause a rift in a relationship, especially when the fortune of one of the friends changes, either up or down. (As we saw between Mercy Otis Warren and Abigail Adams, when the Adamses experienced an upturn in fortune, Mercy Warren became resentful.) Except, perhaps, for the very recent phenomenon of friendships that arise out of social media and a few lasting pen-pal relationships that persist against the odds, economic factors influence most friendships.

### If You Scratch My Back . . .

Throughout history, we have seen women form friendships out of shared concerns, starting with Mary and Elisabeth in the Bible story, who bonded over their pregnancies. The nuns of the Middle Ages shared their faith and physical needs in isolated convents. So, too, did the *salonnières* of the seventeenth century become friends based on their elite social positions and mutual literary pursuits. In early America, women gathered to help one another with household tasks, big and small. Later in the nineteenth century, they would meet to inspire one another to improve their minds, on the one hand, and aid the downtrodden, on the other. The legacy of friendships born in mutual-support groups remains strong today, through any number of civic, self-improvement, and business-advancement groups.

Many women tend to do things for their friends in a proactive way; they often render favors and give small gifts without being asked, and without asking for anything in return. But acknowledged or not, Aristotle's silent code of friendship is invoked: a never-ending round of *manus manum lavat*—one hand washing the other.

Todd May, a present-day political philosopher, looks at relationships in terms of consumerism and entrepreneurship. In his view, every relationship—be it political, social, or personal—has turned into a sort of "market."[293] In this grim assessment of what underlies friendship, May echoes the self-interest theory of La Rochefoucauld in the seventeenth century and later philosophical pessimists.

Consumerism, May suggests, acts like an addictive drug that keeps people from caring about others, since it focuses an individual's attention on fleeting pleasures. Like that of a drug, the consumer "high" fades and has to be renewed periodically.[294] And entrepreneurship encourages individuals to use other people for personal gain.

However, May's concern that market mechanisms are ruining friendships is perhaps overblown. Even if we are sometimes shallow, materialistic, ambitious, and like to shop, that does not mean we are unable to care deeply for a friend, or to cultivate relationships based on affection and trust.

### Friendship in the Workplace

Many women (like men) make friends in the workplace, and these relationships often become vital to their professional lives. A friend at work will watch your back in the maelstrom of office politics. And because she's been in the trenches with you, she can let you vent your work-related stress with an empathy that others in your friendship circle cannot provide.

Today, many people do not distinguish between work and personal friendships. Sheryl Sandberg, chief operating officer of Facebook and a veteran of Google's start-up days, believes that professional relationships should be personal: "I believe in bringing your whole self to work . . . That doesn't mean people have to tell me everything about their personal lives. But I'm pretty sharing of mine." Not all of Sandberg's coworkers are her friends, but one of her long-time associates noted, "The people who are her friends at work are her friends outside work."[295]

Taking a different tack, other arbiters of workplace etiquette recommend that we be cautious, because if a workplace friendship sours, more can go wrong than just the loss of a friend—the awkwardness of daily encounters with someone you would prefer to avoid, for starters. Worse, the possibility of vengeance could result in the loss of a job. Another pitfall lurks in a coworker's inadvertently revealing details of a friend's personal life not meant for general consumption.

Workplace relationships between a supervisor of any rank and her subordinates always require constraints, whether they are a subtle subtext within daily interactions or a rigidly maintained divide. Even when the boss and subordinate are comfortable with their friendship, other coworkers may react with jealousy or paranoia, especially if the subordinate is perceived as sucking up or the supervisor is perceived as playing favorites. Boss/employee friendships lack the equality that Aristotle regarded as a necessary component of true friendship. Perhaps it makes more sense here to substitute the notion of reciprocity for Aristotle's equality. Although two people are never exactly equal in all aspects of their lives, when each brings to the relationship what he or she has to offer in a generous spirit of give and take, then true friendship can sometimes thrive in spite of differences in rank and fortune.

### Outsourcing, Reciprocity, and Friendship

Similarly, friendships sometimes emerge even where one party is a buyer and the other a seller of a personal service, as busy people "outsource" intimate needs that in the past might have been met by friends. Arlie Hochschild noted in *The Outsourced Self: Intimate Life in Market Times*,

> With women in the job force and all Americans working longer hours and having less secure jobs, modern families became ever more hard pressed . . . With no community of yesteryear to lean back on . . . people looked increasingly to the one remaining option—the market.[296]

The market for helping hands now includes caregivers, housekeepers, drivers, handymen, gravesite tenders, personal trainers, cooks, surrogate mothers, rent-a-grandmas, Task-Rabbits (through an online market for performing errands)—you name it. A plethora of online resources has cropped up to connect us to helpful people who will get done what we need done, for pay. Whereas we used to meet people "in the wild," as some online dating sites refer to the old-fashioned way of meeting at a party, we now let others do the vetting and screening. Outsourcing even extends to a market replacement for real friendship: "For those on their own, a pluckily titled Rent-a-Friend service provides a paid 'pal' with whom to eat dinner, see movies, work out at the gym, sort photos, or go on trips—no sex included."[297]

Going to the beauty shop is a familiar way to outsource a formerly personal task. As many a woman and her hairdresser or manicurist know, frequent visits and physical contact provide a rich soil for friendship, be it only what Aristotle would call a friendship of utility. Especially in personal grooming services, an important bond can grow between the groomer and her client. Shelley Taylor cites extensive research on primate grooming behavior, that is, the practice of picking through the pelt of one's companion to remove parasites: "Grooming keeps a friend's coat clean and attractive, and it is also a very soothing, comforting activity."[298] While the modern woman may not want to think of her haircut or manicure as an evolved form of parasite removal, the physical intimacy may be similar. No one today has a handmaiden to perform these services, so we treat ourselves to small luxuries and sometimes become friends with our hair dressers, trainers, or yoga instructors.

Outsourcing is becoming an important conduit for building relationships in our double-duty lives. Leah Busque, CEO of TaskRabbit, found herself amazed at how exponentially her new website grew as people accepted her concept of a "market" between people who need errands done and people who are willing to perform such errands. More than traditional economics was at work. Something like friendship drives many of TaskRabbit's most important transactions. In mid-2014, Leah described this process:

> There's a mom here in San Francisco, and she has a twenty-year-old son living in Boston . . . and unfortunately her son, only twenty years old, was going through chemotherapy treatment . . . She didn't have the money . . . to fly out there to be with him during his treatment. So she went onto TaskRabbit. She found someone [who] could go visit her son in the hospital every day for a week. Bring him a meal, a cozy blanket, sit with him for thirty minutes every single day, and then call her afterwards, give her an update, like, how is he really doing? The person that picked up the job in Boston was actually another mom. And the bond that these two moms formed across the country was incredible. And I realized that what we had built wasn't just about . . . running errands. We were actually redefining who your neighbors are, who you can rely on.[299]

In this case, and in many others, the concept of equality becomes less significant than the concept of reciprocity. If a relationship between two people involves ongoing interactions based in sympathy, empathy, emotional intimacy, mutual caring, and shared values, then a certain kind of friendship will be established, regardless of who is the buyer and who is the seller.

Women today are reinventing their friendships to meet

their needs in a changed society. Many mothers and their nannies find themselves bonding at a deep level, as the nanny is entrusted with nurturance of the most precious thing in the mother's life. Such relationships bear many of the hallmarks of friendship—trust, loyalty, laughing, shared love for a child. Of course, between employer and employee, fair working hours, pay, and benefits must be contractual essentials. But even with these important economic constraints, such a relationship might morph into a mutually caring friendship grounded in something deeper than economics.

Consider the following anecdote[300]: In 2013, in an upscale Northern California community, a typical chatty exchange between a busy professional woman and her once-a-week housecleaner took an unexpected turn. Marisol had been working for Kathy for more than a decade. They enjoyed talking as they went about their respective tasks, and Marisol was pleased to help Kathy learn Spanish. They knew some details about each other's lives—Marisol, by means of cleaning Kathy's house, undoubtedly knew more about Kathy than vice versa, although Kathy showed genuine interest when Marisol discussed her own family and their activities.

One day, Marisol let slip a complaint about her husband: she had two tickets to a sold-out concert in San Jose (Marco Antonio Solís and Lupita D'Alessio) and her husband refused to accompany her. Kathy shared Marisol's dismay at the husband and urged Marisol to take a friend and leave her hubby behind. Marisol said that her spouse would not like her going out without him, but she said she'd take the advice. The next week she reported back with a sigh that none of her girl buddies would go. "What's the matter with them?" Kathy exclaimed indignantly. "Are they nuts? I'd go."

Marisol lit up. "You would? That's great!" Kathy, thinking as a friend, jumped aboard.

The night of the concert, Marisol drove to Kathy's to pick her up. As soon as Kathy came outside in her white-suburbia concert duds—nice jeans, a jacket with a bit of decorative stitching, and boots—Marisol, in a sexy, lamé dress accessorized with strappy spike heels, shook her head in disbelief. She grabbed Kathy's arm and said, "We're going to your closet." Marisol knew the contents of the closet as well as Kathy. There, she dug out an almost-never-worn low-cut top, the one belt with bling that Kathy owned, and Kathy's only spike heels. When Kathy's makeover was finished (including copious additional makeup), Marisol surveyed her work as if she'd made the best of a bad lot and hustled Kathy off to the packed convention center. The place was rocking. Not a word of English was spoken the entire night. Nearly every woman among the tens of thousands of people in attendance was bedecked in glittery party wear with daring décolletage. The music was loud, the sets long and highly danceable. Kathy had a blast.

Was this a good time between friends? Yes. Will Kathy and Marisol become close friends? Probably not. Their different rungs on the socioeconomic ladder will militate against their sharing much of their lives with each other, and the built-in inequality of the employer/employee relationship will always keep an economic barrier between them, even if it is a permeable one. But while the bond between Kathy and Marisol may not be the truest or deepest, it remains an important connection for each of them, which a great many women throughout history and in our busy, pressured lives today gratefully accept as a form of friendship.

Kathy had no problem paying Marisol for her housekeeping service. The economic reality, however, is that the majority of working women can't afford housekeepers or nannies. For nearly every working woman with children, and especially for single mothers, friends are an absolutely necessary backup. Roughly 70 percent of African American households, 50 percent of Hispanic households, and a third of white households are headed by single mothers who know that you can't put kids on hold. Multiple pickups after school, the sick child who must be kept home, the work emergency, the "too-much-to-do-I'm-going-to-lose-it" desperation moment—these are the times a mom calls or texts a friend to toss out the lifesaver and haul her in.

### Ritualized Workplace Friendship: The Mentor

*Mentor* is a buzzword these days, especially among ambitious young women. It's a term that has been much bruited about through the ages, and something has been lost in the translation. As classical scholars know, Mentor was an old man in whose care Odysseus left his son at the onset of the Trojan War. When Athena came to advise the son, Telemachus, she took on the guise of Mentor to avoid raising suspicion among the suitors of Telemachus's mother, Penelope. In disguise, Athena gave the young man encouragement and advice. The modern use of the term "mentor" began around the eighteenth century, with the protagonist of François Fénelon's *Les aventures de Télémaque*. A fundamental aspect of the original story that has been lost to subsequent ages is that Mentor was once an exceedingly powerful goddess.

Mentorship as it applies to modern business undeniably helps junior staff develop the traits necessary for promotion. Even if inequality defines the relationship, the mentor draws on her personal experience to offer good advice with regard to a mentee's problems in the workplace. Is this not similar in many respects to an act of friendship?

The genesis of a rapidly growing young enterprise, Revolution Foods, offers a case in point. Kirstin Groos Richmond, one of the company's two women founders (friends at business school), found herself pregnant in the midst of her hectic start-up. Her mentor provided her with life-changing advice: "I remember Kellie leaning across that desk and saying, 'Whatever you do, do not quit. You can do this. If you go for this and follow your passion, you're actually going to be a better parent . . . you're going to be inspired and fulfilled, and you're going to pass that on to your kids . . .' It's one of those pivotal moments . . . it meant the world to me."[301]

Fortunately, female advancement is a business objective for many leaders in the public and private sectors, and women are beginning to obtain the kinds of mentorship and sponsorship that advance a promising career. Most experts who study executive management distinguish between mentorship and sponsorship. As Sylvia Ann Hewlett of the Center for Talent Innovation notes, "Sponsorship—unlike mentorship, its weaker cousin—makes a measurable difference in career progress. Mentorship, let's be clear, is a relatively loose relationship. Mentors act as a sounding board . . . offering advice as needed and support and guidance as requested; they expect very little in return."[302]

Mentorship does meet many qualifications of what Aristotle called friendships of utility. However, it lacks equality, at least at the outset. The mentor may feel as if she is offering advice to a stand-in for her younger self. That is an attractive indulgence for someone who has made her mark in the world. Over time, as the mentor/mentee relationship deepens, the senior-to-junior dynamic may fall away, leaving not only the trappings of friendship but the real thing.

Sponsorship is far less ambiguous in terms of its relationship to friendship. It is very much a formal business affair. Hewlett describes the no-nonsense tit for tat involved:

To get ahead, women need to acquire a sponsor—a powerfully positioned champion—to help them escape the "marzipan layer," that sticky middle slice of management where so many driven and talented women languish . . . Sponsors, in contrast [to mentors], are much more vested in their protégés, offering guidance and critical feedback because they believe in them . . . Sponsors advocate on their protégés' behalf, connecting them to important players and assignments. In doing so, they make themselves look good. And precisely because sponsors go out on a limb, they expect stellar performance and loyalty.[303]

*Friendship in the Workplace, Third-World Style*

Throughout most of Western history, middle- and lower-class women found it all-consuming to keep body and soul together for their families. Trying to provide food, shelter, and clothing consumed every waking moment. This is still the case for the marginalized poor across the globe.

The World Bank estimated that in 2011, 1.2 billion people, or 17 percent of the world's population, was living in extreme poverty.[304] Of these, 70 percent were women.[305] In developing countries, the economies of villages and urban neighborhoods often center on the work of local women and the friendships among them. Women hold the economic reins of these communities because they take responsibility for the survival of their children. Over the past two decades, this situation has struck a chord with both social entrepreneurs and major lending organizations in the form of microfinance—lending extremely small amounts of money to both urban and rural poor people to help them create, build, and sustain profit-generating enterprises. According to Opportunity International, 93 percent of all microloans are given to women.[306] It has not been lost on microfinance lenders that the best way to avoid default is to extend these loans not to an individual woman but rather to groups of women who trust one another as friends.

Within such a "trust group," each businesswoman exercises her own entrepreneurial vision. One may buy a sewing machine and supplies so she can make clothing from her home while simultaneously minding her small children. Another may weave traditional textiles for sale. They run stores, small farms, beauty shops, and veterinary clinics, among any number of other enterprises. The trust groups provide a built-in monitoring mechanism. The women in the group support one another by sharing business know-how, and they guarantee one another's loans, so no collateral is necessary. If one member runs into difficulty in making a monthly payment, the group understands the reasons—say, an illness in the family, a broken machine, or an inventory imbalance. The group keeps an emergency savings pool to tide members over during such setbacks. The borrower's credit is maintained, her business survives, and the lender stays whole. The grease in this powerful economic engine? Trust among female friends.

### On a Budget: Women Boomers Reinvent Friendship

In recent decades, baby boomer women have tended to outlive men by five to ten years, extending a century-long trend.[307] A large percentage of these older women are single because they have lost their partners to death or divorce, or they never married in the first place. Starting at around age fifty, many single women are deciding to take action to ensure that they can live out their lives in as dignified and happy a manner as possible. For some, this will ultimately mean investing in senior housing. Nearly half a million baby boomer women already live with a roommate, and the trend for such living arrangements appears ready to skyrocket.

General parameters that are emerging for cohousing indicate that women in this age bracket prefer certain areas as communal space, such as the living room and kitchen. Most feel strongly that as long as well-articulated and mutually understood boundaries are in place—for example, who cleans what when, and whether overnight guests are allowed—their households can run peacefully and pleasantly. Very important to most participants in such communal living is well-defined private space.[308] Many lawyers and financial advisors strongly urge even good friends who would be long-term housemates to enter into formal contracts with terms for the myriad unexpected events that can interrupt the most well-intentioned plans. Websites aplenty are cropping up that are designed to help such women make sound decisions about structuring long-term roommate/house-sharing agreements. Issues to be weighed include finance, social interaction, health, and emotional needs.

Such "Golden Girl" arrangements often allow women to live in much nicer neighborhoods than they could afford on their own. Living expenses become a fraction of what running a single-person household would cost. This was the case with Karen Bush, Louise Machinist, and Jean McQuillin, who cowrote *My House Our House*, a lively review of their experiences in forming a shared household in a genteel neighborhood that would have been out of reach for any one of the three alone.[309]

Independent women may surprise themselves by coming to the conclusion that merging their housing resources with a longtime friend makes great sense. Penny, fifty-nine, and Kathy, fifty-two, put this economic principle into action. Bosom buddies since their divorces fifteen years earlier, they sometimes fantasized about becoming roommates, even though they each highly valued privacy. Then, because of a disastrous remodel, Kathy needed a place to stay for several weeks. Penny was more than willing to help her friend, but the remodel stretched out to nine months. The upshot: these two best friends discovered that they made great roommates. Penny explained, "During those nine months we found out we each had as much alone time as we wanted. It just seemed ludicrous to keep paying for two households."[310]

Variations on such arrangements include finding previously unknown roommates from the many online clearinghouses designed for this purpose.[311] Sometimes such arrangements are specifically structured to be transitional for one or both parties. This eases many women into the idea of having a housemate, giving them an out if the other party turns out to be flaky or the roommate from hell. Transitional time in cohousing can also be a godsend for new widows and divorcées, who may find themselves with a lump sum of cash while coping with a multitude of confusing emotions. A reasonable living situation that does not require an immediate outlay gives a woman in transition time to settle down into her new reality and make sound financial decisions.

Marvelously, though not really surprisingly, many such short-term, arm's-length housing arrangements blossom into friendship. Even for highly private people, it can be a great stress reducer to know that someone else will be there should they fall ill or need help during a minor emergency. Margaret Mead is reputed to have said, "One of the oldest human needs is having someone to wonder where you are when you don't come home at night."

Although the trend is just gaining legs, women in this situation are simultaneously reinventing female friendship: "We have become more than friends . . . rather like three sisters who get along, enjoy one another's company, yet go their own ways."[312] Friendship households are becoming another form of ersatz family, as the traditional nuclear model loses its predominance.

While we pretend that the most precious relationships in our lives remain unsullied by economics, our social and financial situations do drive the ways in which we form friendships. The essentials of female friendship have remained constant through the centuries, though the methods for finding and sustaining friends today look considerably different. Women now work far afield from their homes, they are multitasking like never before, and they are finding creative ways to manage the cards life has dealt them.

# FOURTEEN

# CAN WOMEN AND MEN
# BE "JUST FRIENDS"?

*"Affairs happen, and where there's even the possibility of sex, there's gossip."*
—SYLVIA ANN HEWLETT, CENTER FOR TALENT INNOVATION, 2013

*"It's just sex. What's complicated about that? All you're doing is engaging in the most intimate act two people can perform, on a regular basis with someone whom you're attracted to. Things are going to be fine!"*
—GEORGIA WISDOM, "10 RULES FOR FRIENDS WITH BENEFITS," 2014

W HEN HARRY MET SALLY, THE iconic Rob Reiner/ Nora Ephron film starring Billy Crystal and Meg Ryan, resonated with a generation for many reasons, but its central question remains timeless: can women and men be "just" friends? For those who have somehow managed to miss this 1989 cultural watermark, Harry voted no, saying, "The sex thing gets in the way." Yet we have seen that heterosexual friendships form an integral, if complicated, part of the history of women's friendships. Like romantic friendships between two women, relationships between genders fall along a continuum from unambiguous asexuality to active eroticism.

A friend-couple on the platonic end of the scale will find themselves repeating endlessly to others that they are "just friends." The need for such protestations highlights the traditional societal expectation that cross-gender relations involve sex, and that sexual relations are more important than platonic ones. The bias remains today that the assumed relationship between a woman and a man is that of lovers or potential mates, rather than friends. Consider romantic comedies from *Much Ado About Nothing* to the television show, *Friends*, in which the basic premise is that if a couple starts out as friends, they will end up, willy nilly, betrothed or in bed.

Unwieldy cultural or religious baggage from childhood can hamper inter-gender friendship.[313] Fundamentalist sects, in particular, inveigh against contact between men and women other than members of one's family. As one of many Muslim sheikhs who advocates segregating girls from boys put it on YouTube, "A boy cannot have a friendship with a girl unless he has a problem with hormones."[314] More subtle pressure against friendship between boys and girls starts out on many a US playground, with teasing.[315] Later in the lifespan, these barriers grow to include the common obstacles of gossip and spousal jealousy. Gender neutrality in friendship, while increasingly common, is no more a done deal in our culture than is equality in the workplace. Perhaps it is even less so because while equal opportunity laws can be passed and employee ratios and pay scales can be measured, friendship defies objective metrics.

Contemporary social scientists have reported difficulties in determining the nature of nonsexual friendship between women and men. One study suggests that the man in such relationships leaves the door open for eventual sexual involvement (like Harry), while the woman sees the friendship through rosily nonsexual glasses (like Sally, until she changes her mind).[316]

On the other side of the spectrum, Western culture today regards friendship as an important ingredient in successful marriages. This is not an entirely new outlook. Even in classical times, a few philosophers did suppose that a sort of friendship might emerge in marriage. Aristotle, we noted, asserted that husbands and wives naturally coexist in friendship. Montaigne, two millennia later, echoed the notion that a good marriage more closely resembles friendship than love. Companionate marriage became fashionable among the upper classes in eighteenth-century France and England. Today in the United States, equality between marriage partners is generally the ideal. Many a marriage vow declares that the spouses are, above all, best friends. Certainly, married people hope for a shared old age graced with companionship.

### Platonic Friendship

Platonic relationships with men have always been part of the history of women's friendships. To cite but one example, Teresa of Ávila in the sixteenth century wrote of her intense friendship with a priest, who told her about his amorous liaisons with one of his female parishioners. The story fairly oozes sexual tension. Yet Teresa held the line between friendship and romance.

One of the first recorded English uses of *platonic* to refer to an ostensibly nonsexual relationship was the title of the 1631 play *The Platonick Lovers*, written by Sir William Davenant. The protagonists, Eurithea and Theander, begin as platonic friends and intellectual equals. They end as submissive wife and controlling husband. The view that sex cannot be held at bay in woman/man friendships has had its convinced spokesmen from Davenant to Harry.

Nevertheless, modern times do provide considerable opportunity for cross-gender friendships. We expect to find women and men together in our schools, in line at the grocery store, at the next office workstation, in combat zones, and at the neighborhood playground. In all these situations, humans connect with one another. We form tight relationships with colleagues at work—people with whom we may spend more waking hours than we do with our families. We become close with our coed dormmates, teammates, classmates, comrades-in-arms, and fellow parents. As a culture, we may be beginning to outgrow Harry's point of view that friendships between men and women are unrealistic.

### College Daze: Running with the Herd

Young people today, challenging worn-out cultural standards, have created a new American paradigm: packs of girls and boys who intermingle as friends long past childhood, well into their twenties and thirties. Even as the members eventually pair off and marry (much later in life than their parents and grandparents[317]), these contemporary friendship groups continue to include men and women, singles and couples, LGBT and straight people.

For many adults, friendships formed in school and at college remain bedrock bonds. Whether these friends stay in close contact, they function as touchstones for subsequent relationships formed throughout one's life. While cultural biases continue to weigh against women's unimpeded advancement in the workforce, men at institutions of higher learning can no longer easily ignore the competence and potential camaraderie of their equally well-educated women classmates. Today, women outnumber men in undergraduate colleges, medical school, and law school. Consequently, college-attending women have more opportunities than ever to form important friendships with both men and women.

The classroom is not the only venue for high school and college friendships, of course. Two trends have expanded the interaction of girls and boys on campus: increased support for women's athletics, and coed dorms. The former has paved the way for the latter.

### Title IX

Title IX, legislation passed in 1972, bans sex discrimination in any educational program, including sports, that receives federal funding. Given the billions of dollars that are poured into men's football and basketball programs, Title IX meant huge, unprecedented support for girls' high school and college sports. Girls' teams finally got some meaningful funding, and millions of women gained access to the intense and sometimes delightful dynamics of team competition. Few schools, however, had double the funds to replicate boys' sports for girls. To save money, many types of teams in low-contact sports became coed. Especially at the high school level, the coed approach works for teams in sports such as skiing, surfing, swimming, cross-country, track-and-field, triathlons, archery, golf, and karate. Athletes compete in heats against their own gender, but after training and traveling together, girls and boys come to care deeply about the performance of the entire cohort. And they come to care about one another as teammates. They bond by sharing the joy of victory, the pain of loss, the grueling training regimens, and the time hanging out on the sidelines or on the bus as a big, mixed-gender bunch. (In a nod to conventionalism and teenage hormones, locker rooms remain sex-segregated.) Nonathletic extracurricular activities have undergone the same sort of sex-desegregation— think chess club, debate team, robotics club, and so forth.

In terms of women's friendships, Title IX first and foremost gave women an equivalent shot at the important team experience of joint achievements. It allowed great numbers of girls to respect their own and other women's athletic prowess. And it helped change boys' perspectives of girls on the playing field from cheerleaders to teammates. In short, Title IX laid a broad foundation for friendships based in sports that, until its enactment, had been mainly the privilege of men.

With Title IX, not only were sports far more open to girls, but girls and boys on coed teams also gained more chances to know one another as one of the gang, rather than as some mysterious "opposite" creature. The camaraderie of the team easily spilled over into other aspects of life—from the classroom to the boardroom. In the twenty-first century, it has become easier, though still not a slam dunk, to recruit girls into previously male-dominated fields in science, engineering, stand-up comedy, and mathematics. There are plenty of girls, who, given the chance, can design and code with the smartest of the boys. Increasingly, they join traditionally male professions as equals, and potentially find workplace friends of either gender awaiting them.

### Coed Means Boys, Too

Twenty-first-century young women grew up with coed teams and segued into coed dorms in college without batting an eyelash. The transition was not as seamless for those who lived through the sexual revolution. Coed dorms, which showed up in the early 1970s, were at first segregated by floor, with highly vigilant resident administrators stationed in each hallway. Segregated-floor dorms morphed fairly quickly into shared-floor residences, then to shared bathrooms. Today, many campuses allow male and female students to share a room. Most young people generally find this normal, though gender-blind housing remains controversial.[318]

What has happened to all those raging hormones that college administrators have tried to corral? They're still in the air, making campus environs fairly sizzle with sexual energy at times. The upshot is that many girl/boy friendships, after one beer too many, one glorious shared bike ride, or one conversation a bit too intense, slide across the slippery friendship scale, turning their platonic relationship into an erotic one. The outcomes of such scenarios are not foregone conclusions. Young women today prize their friendships—both with women and with men. Thus when friendships turn sexual for young people today, aside from the horrible possibility of rape, there are four likely outcomes: 1) the relationship can change into romance; 2) the young people can awkwardly acknowledge their erotic adventure as a onetime experiment and pick up their friendship where it left off; 3) they can feel that things have become way too close for comfort and stop seeing one another; or 4) they can decide to become friends with benefits, and continue having sex while trying to avoid emotional complications.

### Friends with Benefits

Friends with benefits, a label flung around with increasing frequency, means, according to an online advice column: "It's just sex. What's complicated about that? All you're doing is engaging in the most intimate act two people can perform, on a regular basis with someone whom you're attracted to. Things are going to be fine!"[319]

It appears that in most cases of friends with benefits, friendship trumps sex.[320] The shared wisdom among girls is that sex without romance stops being fun sooner rather than later. One undergraduate described how her best friend, a young man, and she tried out a romantic relationship, but it didn't take, and they were able to revert to being "just" friends. Given their shared passion for long-distance road-biking together, this girl continued to treasure her friendship: "I'm so glad we tried it out. Because now we don't have to wonder about it, and we can just have fun together." This same athlete said that she generally prefers friendships with boys to those with girls because she dislikes the way girls seem to "size me up and look for flaws. They can be so passive-aggressive."[321]

Women are often accused of passive-aggressiveness—the quality that allows one to dig barbs into another while pretending not to. Men's aggressiveness is more often portrayed as direct assault. Whatever the validity of these gender stereotypes, this young woman clearly finds friendly competition with men preferable to the veiled challenges she senses from other women.

On the flip side, some men naturally gravitate toward women as their best friends. When one such guy and his romantic girlfriend of eight years decided to marry, he chose for his best man and groomsman his two best female friends. These "groomswomen" wore different dresses than the bridesmaids, but still in the bride's colors. And they stood on the groom's side of the aisle.

Increasingly, wedding planners are finding that men choose women as their "best man." The opposite is less common. While many women are happy to stand up for their guy buddies, fewer men are comfortable holding a bouquet as a brides-maid equivalent, no matter how dear a friend the bride may be to him. The old double standard appears to be at work in this regard: while it is a "step up" for a woman to assume a man's traditional role, women's roles undertaken by men are still viewed by many as a cultural demotion.

### Cross-Gender Friendship in the Workplace

Having been "acculturated" to friendships with men during their school years, young women today are better prepared to compete in male-dominated fields. At many companies, the men in the new-recruit cohort are more helpful to their women colleagues than the men were in the past. Most sons of mothers who came of age since the feminist revolution have been encouraged to regard women as equals in ways that earlier generations did not. In turn, equality in the workplace eases the way for cross-gender friendship throughout adulthood. Future generations of young women and men may someday be extremely surprised to learn that such friendships were ever considered unusual.

We noted that workplace friendship, especially in the guises of mentorship and sponsorship, can smooth one's path through the professional world. And yet, even well into the twenty-first century, the "sex thing" can get in the way. This happens in a much broader context than Harry contemplated. In recent years, a great deal of media attention has been paid to work-related gender issues. Prevalent among the discussion topics is why women are not rising to the top tiers of management in proportion to their numbers and capabilities. Some emphatic answers cite sexual pitfalls when senior male executives (who at this point constitute the majority of sponsors) take younger women under their professional wing. Sylvia Ann Hewlett of the Center for Talent Innovation states the case:

> Sexual tension will always exist in the workplace. Affairs happen, and where there's even the possibility of sex there's gossip . . . Sex—or the specter of it—haunts sponsorship, prompting men and women to avoid the professional partnerships necessary to achieve their career goals for fear of being censured, fired, or sued.[322]

As more girls and boys grow up in gender-neutral environments, their attitudes toward friendships will carry weight in workplaces of the future. Like most cultural change, this process will likely be a halting progression of the two-steps-forward, one-step-back sort.

### Fag Hags, Lesbros, and Friendships Beyond Labels

During the last decades of the twentieth century, the derogatory term *fag hag* referred to a girl or woman who generally preferred hanging out with gay men instead of with women. The coinage of the term reflected the growing prevalence of such friendships—at least through the coarse lens of the public eye. What most people failed to realize was that such friendships were nothing new. For ages, many women had been best friends with gay men, whether or not they consciously recognized their friends' sexual orientation. But in the 1970s and 1980s, many more gay people came out of the closet and declared their sexual identity to the world at large. Women who would have been close friends with these men in any case could now relate to them in a more honest, open way.

By the 1990s, gay men and their friends defused the hurtful implications of the fag hag descriptor by ironically embracing it. Similarly, in a gender-equality twist on women's friendships, "lesbros" have joined the denizens of lesbian bars. These men enjoy the company of their lesbian friends, who often join them at typically male venues as well. All this mixing of sexual identity across friendship lines has rapidly become overused on TV, especially after sitcoms like *Will and Grace* and *Sex in the City* depicted attractive straight women as close friends with attractive gay men. Absurdly bowdlerized, such shows made gay people more palatable to mainstream America. Like most sitcoms, the shows eventually lost their cachet, and so have *fag hag* and other ironic labels used for relationships between straight people and LGBT individuals. What once was a socially outré relationship became a cliché. In 2009, Thomas Rogers, Salon.com's former arts editor, slammed some nails into the fag hag coffin:

> Nowadays, when a grown woman describes herself as a "fag hag," it feels like she's throwing around a designer label or telling me she knows a celebrity—a kind of social conspicuous consumption . . . So what happens to those fabulous gay-loving straight women of yore? . . . Hopefully

they'll be calling themselves something more accurate. Like "friend."[323]

### The Divorcée

Through much of modern history, middle-class married women formed friendships through their husbands' social status and connections. When we looked at couple friendships in the mid–twentieth century, a husband's friends— and the wives of his friends—were supposed to be his wife's friends as well. Naturally, there were various cross-couple arrangements, depending on personalities. If the women in such couple-to-couple friendships were lucky, they would hit it off.

The wheels came off the two-couple friendship wagon in the late 1960s and 1970s, in part due to the free-love era, which encouraged looser sexual mores. As cultural norms shifted, some foursomes of couple friends found themselves in bed together. This mate-swapping trend, captured in the film *Bob & Carol & Ted & Alice* (1969), along with a newfound propensity toward marital infidelity, blasted the old paradigm of the friendly couples set. For a variety of complex reasons, marriage has never regained its place as a stable social structure. Today, 50 percent of first marriages fail, a statistic that has powerful implications for women's friendships.[324]

When couples divorce, they cast their mutual friends into a sea of awkwardness. It's unthinkable not to invite your best friends to a party, but now that they're divorced, whom do you invite? The ex-wife? The ex-husband? Both of them? And if each half of the split couple remarries, are you supposed to substitute the new-mate couple for your displaced friend, just like that? Or, worse, put the new-mate couples in the same room?

Many divorced women draw the short straw on social-group reconfigurations post-breakup. Even today, the use of *divorcée* to describe a woman carries a slight frisson of disapproval. (And the male-counterpart term, *divorcé*, is almost never used.) For women who have focused their friendships around their marriage, this can be devastating.

Fortunately, in the twenty-first century, there are new friendship options for divorced women. Many online meetup groups are full of people looking for companionability around an activity, without being on the prowl for a mate. And while nobody likes to think that a marriage won't last, the high divorce rate does mean that the solace of friendship with people in similar circumstances is ever more available.

### Make New Friends, But Keep the Old / One Is Silver and the Other Gold

While the permanence of marriage as an institution is questionable, the value placed on lifelong friends has not waned. Many middle-age people are discovering that those women/men herds of their youth want to reunite and kick up their heels every now and again. Baby boomers who lived on the front lines of mixed-gender dorms in the 1970s and 1980s easily maintain friendships among the whole gang with the aid of the Internet. One woman graduate of 1988 wrote about her coed dormmates, still vital in her friendship network:

> Our friendships are easily renewed as time passes, whether at a memorial service, . . . homecoming, or on Facebook, where we continue to share our varied lives . . . So thanks, housing office . . . for two floors of men and one of women, for so many tears and for so, so many laughs. Friends are the family we choose."[325]

Cross-gender friendship in America is unmistakably on the rise. That said, Harry did have a point. Sex complicates matters of friendship, often changing friendly relationships into categorically erotic ones. However, one of the sublime beauties of true friendship is that it need not be constrained by language. Why use the term "just friends" when we can simply call a friend a friend?

## FEMALE FRIENDSHIP:

## WHAT ENDURES

B Y ADOPTING A CULTURAL-HISTORICAL APPROACH to female friendship in the Western world, we have witnessed a remarkable evolution from near invisibility to iconic prominence. For more than two millennia, from circa 600 BCE to circa 1600 CE, women's friendships were ignored, belittled, or openly denigrated by male creators of the written record. However, for the past four hundred years, as women have gained access to literacy, socioeconomic resources, and civil rights, their friendships have become increasingly visible, so much so that women in the twenty-first century are now setting the model of friendship for both sexes.

What is it about women as friends that is so special? What has endured through the centuries in different settings, languages, and cultures? Are there qualities of women's friendship that seem to be almost universal, despite the various outward forms that friends adopt in one milieu or another? What common elements can be found in the friendships between two sixteenth-century English "gossips," two seventeenth-century French aristocrats, two eighteenth-century elite Americans, two nineteenth-century German or American sweethearts, two twentieth-century African American or feminist "sisters," and two twenty-first-century working women? From the many examples presented in this book, we have identified the following four ingredients that seem basic to women's friendships.

- **Affection**. Women's friendships with one another have an emotional core that ranges in intensity from empathy and goodwill to passion and love. The word *affection* seems to apply to every case we have studied. Without affection—defined as a kindly feeling or fondness for the other person— there is no such thing as women's friendship.

- **Self-revelation**. A woman's friend is someone she can talk to openly, without fear of reprisal and with the expectation of sympathy and support. Yes, women friends talk—they gossip; they confide in each other; they tell each other things they wouldn't want their parents, spouses, or children to know. Long ago, a man complained bitterly to one of the authors of this book: "Women tell each other *everything*!" Today he might be more prone to ask himself why he has so few friends at the age of sixty-five.

• **Physical contact**. Women friends touch each other, they hug, they kiss (though not usually on the mouth); they rub each other's backs, wash each other's hair, paint each other's nails. They exchange clothes and dress each other up. They take care of the sick and dying. Whatever the reasons—nature or nurture—female friends are usually given to a far broader range of physical contact than male friends. Men do give each other hugs and slaps on the back (or, in sports, on the butt), and gay couples sometimes walk hand-in-hand in tolerant venues, but in the United States, straight men usually refrain from the enthusiastic embraces that are characteristic of women.

• **Interdependence**. Women depend on one another from the time they are girls to the time they are working women, mothers, divorcées, or widows. Schoolgirls ask each other how to dress, how to attract a desired boyfriend, and how to prepare for the next exam. Working women count on their friends to help identify a better job situation or how to negotiate a pay raise, not to mention how to deal with sexual harassment. Mothers count on one another for backup—to pick up a child at school or drive half a team to their soccer game. Women going through a divorce turn to their friends for sympathy and support. Single women, divorcées, and widows meet regularly for tennis, bicycling, and book clubs; go on vacations together; and sometimes end up living together in duos or group arrangements. Women traditionally have bonded together to help one another cope and survive.

### The Future of Friendship

We have followed a zigzag path of female friendship through Western history up to present-day America. Friendship, defined as a personal relationship grounded in mutual affection, empathy, reciprocity, and support, has become a prized staple of American life today, especially for women. In contrast to marriage, friendship has no legal underpinnings, no economic obligations, no children to raise, no "for better, for worse" promises. Yet while many American marriages founder on unexpected shoals, such as all-consuming jobs, diminished sexual energy, uncontrollable children, and financial reverses, friendships are flourishing as never before. American society is beginning to pull away from the notion that a marital partner can fulfill all a person's needs. Friends step in to fill the breach, before, during, and after marriage. They may be the saving grace in lives that are overworked, overstressed, and less committed to extended family.

We foresee a future for friendship that will incorporate, for American men as well as for women, at least three of the aspects identified as "feminine": affection, self-disclosure, and interdependence. These characteristics are already seeping into public debate, with men asking whether it is necessary to suppress their emotions and button their lips in order to appear manly. Taking a cue from their wives, their women co-workers, and their women friends, some men have found that sharing more personal information about themselves with their buddies can incite responses that are valuable to their relationships and careers. In the past, men may have maintained power by projecting an image of tough self-sufficiency, but today it is not uncommon for some—perhaps many—men to express their feelings more openly and to admit their dependence on their friends and colleagues, as well as on their wives.

To think of friendship as a self-revealing, emotional connection between two (or more) people veers away from older, masculine notions that emphasized comradeship, solidarity, and citizenship. The vision of men standing side by side as soldiers or lining up as members of the same team has not disappeared, but it has opened out to include the picture of men—like women—looking at each other face-to-face, verbally expressing their innermost concerns, and hugging each other as a sign of affection.[326]

Just as men have "feminized" some aspects of their friendships, so, too, have women taken on military and civic roles that once belonged only to men. They now stand side by side as comrades-in-arms with other men and women and sit alongside their colleagues in the Senate and House of Representatives. Aristotle's vision of a society based on friendship between men has evolved into a society girded by both genders, with women as well as men responsible for the civic good.

Both women and men have entered a new arena in which mixed-gender friendships are becoming more and more common. In schools, colleges, and the workplace; in churches, clubs, and service organizations; and on websites and at meetups, men and women are now intermingling with an ease that would have shocked their Victorian ancestors. Girlfriends, boyfriends, and just friends constitute the new frontier where Americans are trying to create rewarding relationships beyond the boundaries of gendered stereotypes.

Our history suggests that women will continue to show the world how to be friends. With the uncertainties surrounding marriage, it is likely that friendships will continue to offer forms of support women would once have found within their families. We can expect to see more single women living together as roommates, and more older women sharing households. In our utopian fantasies, we imagine a world in which the strengths of the friendly sex imbue society with greater concern for the well-being of every person.

# NOTES

1. Shelley E. Taylor, *The Tending Instinct: How Nurturing Is Essential to Who We Are and How We Live* (New York: Henry Holt, 2002), 90.

2. Dr. Louann Brizendine (author of *The Female Brain* and *The Male Brain*), personal communication with the author, 2013.

3. For example, Peter M. Nardi, ed., *Men's Friendships* (London: Sage Publications), 1992.

4. Geoffrey L. Greif, *Buddy System: Understanding Male Friendships* (New York: Oxford University Press, 2009), 6.

5. Pauline Nestor, *Female Friendships and Communities: Charlotte Brontë, George Eliot, Elizabeth Gaskell* (New York: Oxford University Press, 1985), 4.

6. Gertrude Franklin Horn Atherton, *The Conqueror: Being the True and Romantic Story of Alexander Hamilton* (New York: The MacMillan Company, 1904), 231.

7. C. S. Lewis, *The Four Loves* (Orlando, FL: Harcourt Books, 1960).

8. Leon Battista Alberti, "On the Family," quoted in Julia O'Faolain and Lauro Martines, eds., *Not in God's Image: Women in History from the Greeks to the Victorians* (New York: Harper & Row, 1973), 189.

9. Michel de Montaigne, *The Complete Essays of Montaigne*, trans. Donald M. Frame (Stanford, CA: Stanford University Press, 1965), 138.

10. Quoted in Carolyn James and Bill Kent, "Renaissance Friendships: Traditional Truths, New and Dissenting Voices," in *Friendship: A History*, ed. Barbara Caine (Oakville, Ontario: Equinox, 2009), 149.

11. Edith Gelles, chapter three in *"First Thoughts": Life and Letters of Abigail Adams* (New York: Twayne, 1998).

12. "Ngram Viewer," Google Books books.google.com/ngrams; Jean-Baptiste Michel et al., "Quantitative Analysis of Culture Using Millions of Digitized Books," *Science* online, December 16, 2010.

13. Job 2:13 (King James Bible), kingjamesbibleonline.org. Hereafter cited in text.

14. Willis Barnstone, trans., "Miryam of Magdala," in *The Restored New Testament* (New York: Norton, 2009), 583.

15. Elaine Pagels, *The Gnostic Gospels* (New York: Random House, 1979); Karen King, *The*

*Gospel of Mary of Magdala: Jesus and the First Woman Apostle* (Santa Rosa, CA: Polebridge Press, 2003).

16. For examples of the most recent, Todd May, "Friendship in an Age of Economics," *New York Times*, July 4, 2010; Ray Pahl, *On Friendship* (Cambridge, UK: Polity Press, 2000).

17. David Konstan, *Friendship in the Classical World* (Cambridge, UK: Cambridge University Press, 1977), 1.

18. Eva Österberg, *Friendship and Love, Ethics and Politics: Studies in Mediaeval and Early Modern History* (New York: Central European University Press, 2010), 26.

19. This and the preceding quotations are all from Aristotle, Books VIII and IX in *Nicomachean Ethics*, in *Introduction to Aristotle*, ed. Richard McKeon (New York: The Modern Library, 1947).

20. Authorized doctrine 27, quoted in Norman Wentworth DeWitt, *Epicurus and His Philosophy* (Minneapolis: University of Minnesota Press, 1954), 190.

21. Vatican Saying 23, ibid., 308.

22. Vatican Saying 39, ibid.

23. Plutarch, *Oeuvres Morales* (Paris: Société d'Édition "Les Belles Lettres," 1985), 2:152. Translation from the French by Marilyn Yalom.

24. This and the preceding Cicero quotations are from Cicero, *On Old Age and On Friendship*, trans. Frank O. Copley (Ann Arbor: University of Michigan Press, 1967), 45–90.

25. Constant J. Mews, "Cicero on Friendship," in *Friendship: A History*, ed. Barbara Caine, 67–71.

26. Introduction to *Friendship in the Middle Ages and Early Modern Age*, ed. Albrecht Classen and Marilyn Sandidge (Göttingen: De Gruyter, 2010), 11.

27. This and the preceding quotations are from Maria Boulding, trans., *The Works of Saint Augustine: The Confessions* (Hyde Park, NY: New City Press, 1997), Part I, vol. 1, Book IV:7, 96–100.

28. Christoph Cardinal Schönborn, archbishop of Vienna (commencement address, Thomas Aquinas College, California, June 8, 2002).

29. Walter Frölich, trans., *The Letters of Saint Anselm* (Kalamazoo, MI: Cistercian Publications, 1990), 1:285.

30. Ibid., 1:210.

31. Ibid., 1:81.

32. Brian Patrick McGuire, *Friendship and Community: The Monastic Experience 350–1250* (Ithaca: Cornell University Press, 2010), 214.

33. Ibid., 194.

34. George Lawless, ed., *Augustine of Hippo and His Monastic Rule* (Oxford: Oxford University Press, 1987), 81.

35. Jo Ann Kay McNamara, *Sisters in Arms: Catholic Nuns through Two Millennia* (Cambridge, MA: Harvard University Press, 1996), 76.

36. Bernadette J. Brooten, *Love Between Women: Early Christian Responses to Female Homoeroticism* (Chicago: University of Chicago Press, 1996), 350–51.

37. Silvia Evangelisti, *Nuns: A History of Convent Life 1450–1700* (Oxford: Oxford University Press, 2007), 30–31.

38. Philippe de Navarre, *Les Quatre Ages de l'Homme* (Paris: Librairie Firmin Didot, 1888), 16.

39. Fiona Maddocks, *Hildegard of Bingen: The Woman of Her Age* (New York: Doubleday, 2001), 17–24.

40. *Jutta and Hildegard: The Biographical Sources,* ed. Anna Silvas (University Park, PA: The Pennsylvania State University Press, 1999), 165.

41. Letter 12 in *The Letters of Hildegard of Bingen,* trans. Joseph L. Baird and Radd K. Ehrman (New York: Oxford University Press, 1994), 1:48.

42. Letter 64, ibid., 143–44.

43. Letter 13, ibid., 50.

44. Julie Ann Smith, *Ordering Women's Lives: Penitentials and Nunnery Rules in the Early Medieval West* (Burlington, VT: Ashgate, 2001), 191–92.

45. M. Colman O'Dell, "Elisabeth of Schönau and Hildegard of Bingen: Prophets of the Lord," in *Peace Weavers: Medieval Religious Women,* ed. Lillian Thomas Shank and John A. Nichols (Kalamazoo, MI: Cistercian Publications, 1987), 2:88.

46. Letter 201r, *Letters of Hildegard,* 2:181.

47. Letter 100, ibid.,15.

48. Letter 157, ibid., 104.

49. Letters 140 and 140r, ibid., 80.

50. Letter 150, ibid., 95.

51. Mary Jeremy Finnegan, *The Women of Helfta: Scholars and Mystics* (Athens, GA: University of Georgia Press, 1991), 27.

52. Ann Marie Caron, "Taste and See the Goodness of the Lord: Mechtild of Hackeborn," in Book Two of *Hidden Springs: Cistercian Monastic Women,* ed. John A. Nichols and Lillian Thomas Shank (Kansas City, MO: Cistercian Publications, 1995), 509–24.

53. Evangelisti, *Nuns,* 72.

54. Ibid., 80–81.

55. Evangelisti, *Nuns,* 60.

56. Penelope D. Johnson, *Equal in Monastic Professions: Religious Women in Medieval France* (Chicago: University of Chicago Press, 1991), 121.

57. Judith C. Brown, *Immodest Acts: The Life of a Lesbian Nun in Renaissance Italy* (New York: Oxford University Press, 1986), 4.

58. Ibid., 117–18.

59. Kieran Kavanaugh, OCD, and Otilio Rodriguez, OCD, trans., "The Book of Her Life," in *The Collected Works of St. Teresa of Ávila* (Washington, DC: ICS Publications, 2012), 1:95.

60. Ibid., 70.

61. Ibid., 71.

62. Ibid., 72–73.

63. Ibid., 211–12.

64. Alison Weber, "'Little Angels': Young Girls in Discalced Carmelite Convents (1562–1582)," in *Female Monasticism in Early Modern Europe,* ed. Cordula van Wyhe (Burlington, VT: Ashgate, 2008), 212.

65. Evangelisti, *Nuns,* 76; Alison Weber and Amanda Powell, eds., *Book for the Hour of Recreation* (Chicago: University of Chicago Press, 2002), 43.

66. Ernest W. McDonnell, *The Beguines and Beghards in Medieval Culture: With Special Emphasis on the Belgian Scene* (New Brunswick, NJ: Rutgers University Press, 1954); Marguerite Porete, *The Mirror of Simple Souls,* trans. Ellen L. Babinsky (Mahwah, NJ: Paulist Press,

1993); Dennis Devlin, "Feminine Lay Piety in the High Middle Ages: The Beguines," in *Distant Echoes: Medieval Religious Women*, ed. John Al Nichols and Lillian Thomas Shank (Kalamazoo, MI: Cistercian Publication), vol. 1.

67. Herbert Grundmann, *Religiöse Bewegungen im Mittelalter* (Berlin, 1935; Hildesheim, 1961). Citation is to the Hildesheim edition.

68. Octavio Paz, *Sor Juana*, trans. Margaret Sayers Peden (Cambridge, MA: Harvard University Press, 1988), 90.

69. Sor Juana Inés de la Cruz, *The Answer/La Respuesta: Including a Selection of Poems*, trans. Electa Arenal and Amanda Powell (New York: The Feminist Press at the City University of New York, 1994), 11; see also Sor Juana Inés de la Cruz, *Obras Completas*, ed. Alfonso Méndez Plancarte (Mexico: Fondo de Cultura Económica), 1:240–42.

70. Sor Juana Inés de la Cruz, *Redondilla 91, Selected Works,* trans. Edith Grossman (New York: W.W. Norton, 2014), 30.

71. Emanuel van Meteren, *Album*, quoted in Germaine Greer, *Shakespeare's Wife* (New York: HarperCollins, 2007), 30.

72. Melinda Jay, "Female Friendship Alliances in Shakespeare" (doctoral dissertation, Florida State University, 2008).

73. William Shakespeare, *As You Like It*, 1.3.69–72. *The Comedies of Shakespeare*, Vol. Two, (New York: The Modern Library, 1959)

74. Peter Ackroyd, *Shakespeare: The Biography* (London: Chatto and Windus, 2005), 29.

75. Sara Mendelson and Patricia Crawford, *Women in Early Modern England, 1550–1720* (Oxford: Oxford University Press, 1998), 240–42.

76. Ann Rosalind Jones, "Maidservants of London: Sisterhoods of Kinship and Labor," in *Maids and Mistresses, Cousins and Queens: Women's Alliances in Early Modern England*, ed. Susan Frye and Karen Robertson (New York: Oxford University Press, 1999), 21–23.

77. Greer, *Shakespeare's Wife*, 129.

78. Antonia Fraser, *The Weaker Vessel* (New York, Alfred A. Knopf, 1984), 337; Germaine Greer and others, eds., *Kissing the Rod: An Anthology of Seventeenth-Century Women's Verse* (New York: The Noonday Press, 1988), 186.

79. Katherine Philips, "L'Amitié," in Greer, *Kissing the Rod*, 189–90. Philips's poems can also be found online at Luminarium.org.

80. Katherine Philips, "On Rosania's Apostasy and Lucasia's Friendship," in Greer, *Kissing the Rod*, 194–95.

81. Katherine Philips, "Friendship's Mysterys, to my dearest Lucasia," in Greer, *Kissing the Rod*, 193.

82. Katherine Philips, "To My Excellent Lucasia, On Our Friendship," in *Poems Between Women: Four Centuries of Love, Romantic Friendship, and Desire*, ed. Emma Donaghue (New York: Columbia University Press, 1997), 3.

83. Quoted in Fraser, *The Weaker Vessel*, 338.

84. Quoted in Linda W. Rosenzweig, *Another Self: Middle-Class American Women and Their Friends in the Twentieth Century* (New York: New York University Press, 1999), 21.

85. Jane Harrison, *Reminiscences of a Student's Life* (London: Hogarth Press, 1925).

86. Valerie Traub, "'Friendship So Curst': Amor Impossibilis, the Homoerotic Lament and the Nature of Lesbian Desire," in *Lesbian Dames: Sapphism in the Long Eighteenth Century*, ed. John C. Beynon and Caroline Gonda (Burlington, VT: Ashgate Publishing, 2010),

10. Traub's chapter offers valuable insights into Philips, her cohort, and friendship as a literary topos.

87. Quoted in Rosenzweig, *Another Self*, 15.

88. Beynon and Gonda, *Lesbian Dames*, 30–34.

89. Elizabeth Robinson Montagu to Sarah Robinson, 18 Sept. 1750, quoted in Susan S. Lanser, "Tory Lesbians: Economies of Intimacy and the Status of Desire," in Beynon and Gonda, *Lesbian Dames*, 173.

90. Laurel Thatcher Ulrich, *Good Wives: Image and Reality in the Lives of Women in Northern New England 1650–1750* (New York: Knopf, 1982), 9.

91. Ibid., 121–22.

92. Quoted in Georges Mongrédian, *Les Précieux et les Précieuses* (Paris: Mercure de France, 1963), 72–84. All translations from the French unless otherwise noted are by Marilyn Yalom.

93. Madeleine de Scudéry, *Artamène ou le Grand Cyrus*, quoted in ibid., 119–22.

94. Myriam Maître, *Les précieuses. Naissance des femmes de lettres en France au XVIIe siècle* (Paris: Honoré Champion, 1999), 281.

95. Madeleine de Scudéry, *Artamène ou le Grand Cyrus*, quoted in Mongrédian, *Les Précieux et les Précieuses*, 121–22.

96. Madeleine de Scudéry, *Clélie*, quoted and discussed in Roger Duchêne, *Les Précieuses ou comment l'esprit vint aux femmes* (Paris: Fayard, 2001), 30–31.

97. Jacqueline Quenau and Jean-Yves Patte, *L'Art de vivre au temps de Madame de Sévigné* (Paris: NiL editions, 1996), 160.

98. La Fayette to Sévigné 24 Jan. 1692, *Œuvres Complètes* (Paris: La Pléiade, 2014), 1082.

99. Denise Mayer, *Une Amitié parisienne au Grand siècle: Mme de Lafayette et Mme de Sévigné, 1648–1693* (Seattle: Papers on French Seventeenth-Century Literature/Biblio 17, 1990), 46–47.

100. La Fayette to Ménage, Sunday evening, Aug. 1660, in La Fayette, *Œuvres Complètes*, 921.

101. Quoted in Quenau and Patte, *L'Art de vivre*, 208.

102. La Fayette to Ménage, 17 July 1657, in La Fayette, *Œuvres Complètes*, 902.

103. *Divers Portraits* (Caen, 1659). This collection of portraits was written by various hands.

104. This and the preceding quotations are in La Fayette, *Œuvres Complètes*, 3-5.

105. Sévigné to the Marquis de Pomponne, 1 Aug. 1667, in Madame de Sévigné, *Correspondance*, 1 (Paris: La Pléiade, 1972), 87.

106. Sévigné to Grignan, 8 July 1672, ibid., 550.

107. La Fayette to Sévigné, 14 July 1673, in La Fayette, *Œuvres Complètes*, 983.

108. La Fayette to Sévigné, 4 Sept. 1673, in ibid., 983.

109. La Fayette to Sévigné, ibid.

110. Mayer, *Une Amitié parisienne*, 93–94.

111. Sévigné to Grignan, 6 March 1680, in Mme de Sévigné, *Correspondance*, 2 (Paris: La Pléiade, 1974), 860.

112. Sévigné to Grignan, 15 March 1680, in ibid., 875.

113. Sévigné to Grignan, 17 March 1680, in ibid., 876.

114. La Fayette to Ménage, c. May 1684, in La Fayette, *Œuvres Complètes*, 1027.

115. La Fayette to Ménage, Sept. 1691, in inbid., 1057.

116. Sévigné to Grignan, 26 Feb. 1690, in Mme de Sévigné, *Correspondance*, 3 (Paris: La Pléiade, 1978), 847.

117. Sévigné to Guitaut, 3 June 1693, in ibid., 1006.

118. Marilyn Yalom, chapter one in *How the French Invented Love: Nine Hundred Years of Passion and Romance* (New York: HarperCollins, 2012).

119. Linder Kerber, "The Republican Mother and the Woman Citizen: Contradictions and Choices in Revolutionary America," in *Women's America, Refocusing the Past*, ed. Linda Kerber and Jane Sherron De Hart (New York: Oxford University Press, 2000), 112–20.

120. Edith Gelles, *"First Thoughts": Life and Letters of Abigail Adams* (New York: Twayne Publishers, 1998), 35.

121. Abigail to Mercy, December 1773, quoted in Edith Gelles, *Abigail and John: Portrait of a Marriage* (New York: William Morrow, 2009), 39.

122. Caroline Winterer, *The Mirror of Antiquity* (Ithaca: Cornell University Press, 2007), 8.

123. Gelles, *"First Thoughts,"* 47–48.

124. Ibid., 14–18.

125. Ibid., 51.

126. Kate Davies, *Catherine Macaulay and Mercy Otis Warren: The Revolutionary Atlantic and the Politics of Gender* (Oxford: Oxford University Press, 2005), 195. See also Edith Gelles, *Portia: The World of Abigail Adams* (Bloomington, IA: Indiana University Press, 1992), 3–4.

127. Gelles, *"First Thoughts,"* 60.

128. Ibid., 168.

129. Ibid., 60.

130. Davies, *Catherine Macaulay and Mercy Otis Warren*, 2.

131. Catherine Macaulay, "An Address to the People of England, Scotland and Ireland on the Present Important Crisis of Affairs," Dec. 1774, quoted in ibid., 1.

132. Warren to Macaulay, Dec. 1774, quoted in *Catherine Macaulay and Mercy Otis Warren*, 1.

133. Macaulay to Warren, 15 July 1785, quoted in ibid., 20.

134. See, for example, Darline Levy, Harriet Applewhite, and Mary Johnson, eds., *Women in Revolutionary Paris, 1789–1795* (Urbana: University of Illinois Press, 1980); Anne Soprani, *La Révolution et les Femmes de 1789 a 1796* (Paris: MA Editions, 1988); Marilyn Yalom, *Blood Sisters: The French Revolution in Women's Memory* (New York: Basic Books, 1993).

135. Levy, Applewhite, and Johnson, *Women in Revolutionary Paris*, 215.

136. "Les Souvenirs de Sophie Grandchamp," in appendix to *Mémoires de Madame Roland*, ed. Claude Perroud (Paris: Plon-Nourrit et Cie, 1905), 2:461–97.

137. Claude Perroud, *Mémoires de Madame Roland* (Paris: Mercure de France, 1987), 155.

138. This and previous quote "Les Souvenirs de Sophie Grandchamp," ibid., 2:492–95.

139. Nancy Cott, *The Bonds of Womanhood: "Woman's Sphere" in New England, 1780–1835* (New Haven: Yale University Press, 1977, repr. 1997), 160.

140. Alison Oram and Annmarie Turnbull, eds., *The Lesbian History Sourcebook: Love and Sex Between Women in Britain from 1780 to 1970* (London: Routledge, 2001), 55.

141. Anna Seward, *Llangollen Vale, with Other Poems* (London: G. Sael, 1796; Open Library, 2009), 6, https://openlibrary.org/works/OL2067573W/Llangollen_vale_with_other_poems.

142. John D'Emilio and Estelle Freedman, *Intimate Matters: A History of Sexuality in America* (New York: Harper & Row, 1988), 192–93.

143. Carroll Smith-Rosenberg, "The Female World of Love and Ritual," in *Disorderly Conduct: Visions of Gender in Victorian America* (New York: Oxford University Press, 1985), 76.

144. Angele Steidele, *Geschichte einer Liebe: Adele Schopenhauer und Sibylle Mertens* (Berlin: Insel Verlag, 2010). All translations by Marilyn Yalom.

145. Anna Jameson, *Visits and Sketches at Home and Abroad* (New York: Harper & Brothers, 1834), 1:36.

146. Adele to Ottilie, 1814, in Steidele, *Geschichte einer Liebe*, 44.

147. Ottilie to Adele, July 1814, in Steidele, *Geschichte einer Liebe*, 45.

148. Ibid., 63.

149. Adele to Ottilie, 8 June 1828, in Steidele, *Geschichte einer Liebe*, 80.

150. Adele to Ottilie, 1 July 1829, in Steidele, *Geschichte einer Liebe*, 86–87.

151. Jameson, *Visits and Sketches*, 1:36.

152. Steidele, *Geschichte einer Liebe*, 101.

153. Sibylle to Adele, 8 March 1836, ibid., 145–46.

154. Fanny Lewald, *Römisches Tagebuch 1845/46*, ed. Heinrich Spiero (Leipzig: Klinkhardt & Biermann, 1927), 57, quoted in Steidele, *Geschichte einer Liebe*, 215.

155. Anna to Ottilie, 9 August 1845, in Steidele, *Geschichte einer Liebe*, 214.

156. George Sand, part III, chapters xii–xiii, in *Story of My Life: The Autobiography of George Sand*, ed. Thelma Jurgrau (Albany: State University of New York Press, 1991).

157. Ibid., 685.

158. Ibid., 684.

159. Ibid., 689.

160. Ibid., 685.

161. Erna Olafson Hellerstein, Leslie Parker Hume, and Karen M. Offen, eds., *Victorian Women: A Documentary Account of Women's Lives in Nineteenth-Century England, France, and the United States* (Stanford, CA: Stanford University Press, 1981), 89.

162. Dorothy Wordsworth, "Irregular Verses," in *Poems Between Women: Four Centuries of Love, Romantic Friendship, and Desire*, ed. Emma Donaghue (New York: Columbia University Press, 1997), 38–41.

163. Christina Rossetti, "Gone Before," in Donaghue, *Poems Between Women*, 65–66.

164. Frances Osgood, "The Garden of Friendship," in Donaghue, *Poems Between Women*, 53–54.

165. Anya Jabour, *Scarlett's Sisters: Young Women in the Old South* (Chapel Hill, NC: University of North Carolina Press, 2007), 70–76.

166. Ibid., 73.

167. Ibid., 79.

168. Ibid., 71.

169. William R. Taylor and Christopher Lasch, "Two 'Kindred Spirits': Sorority and Family in New England, 1839–1846," in *History of Women in the United States, Vol. 16, Women Together: Organizational Life*, ed. Nancy F. Cott (New Providence: K. G. Saur, 1994), 93.

170. Case to Edgarton, 18 Oct. 1839, in ibid., 85.

171. Edgarton to Case, 8 Jan. 1840, in ibid., 94.

172. Edgarton to Case, in ibid., 98.

173. Lillian Faderman, "Nineteenth-Century Boston Marriage as a Possible Lesson for Today," in *Boston Marriages: Romantic but Asexual Relationships Among Contemporary Lesbians*, ed. Esther D. Rothblum and Kathleen A. Brehony (Amherst, MA: University of Massachusetts Press, 1993), 32.

174. Helena Whitbread, ed., *No Priest but Love: The Journals of Anne Lister from 1824–1826* (Ottley, UK: Smith Settle, 1992); Helena Whitbread, ed., *The Secret Diaries of Miss Anne Lister (1791–1840)* (London: Virago, 2010).

175. Sunday, 6 Jan. 1822 in Whitbread, *The Secret Diaries*, ibid., 194.

176. Thursday, 16 March 1826 in Whitbread, *No Priest but Love*, 163.

177. 8 Jan. 1834; 10 Feb. 1834; 27 Feb. 1834; 23 May 1834, in Jill Liddington, *Female Fortune: Land, Gender and Authority, The Anne Lister Diaries and Other Writings, 1833–36* (New York: Rivers Oram Press, 1998), 86, 92, 95, 107.

178. Ibid., 242.

179. Eliza Schlatter to Sophie Dupont, 24 Aug. 1834, in Carroll Smith-Rosenberg, *Disorderly Conduct*, 73.

180. The letters quoted in this chapter are from the Mary Hallock Foote Papers, MO115, Dept. of Special Collections, Stanford University Libraries, Boxes 1-4.

181. Ibid., Box 1, 8 Feb. 1870.

182. Ibid., 28 Sep. 1873.

183. All of this has provoked lively debate among scholars: Lillian Faderman and Esther Rothblum take the position that many romantic friendships in the past were nonsexual and that many still are, whereas Terry Castle insists on the carnal specificity of lesbian desire, "its incorrigibly lascivious surge toward the body of another woman," not only in the present but also in the past. Lillian Faderman, *Surpassing the Love of Men: Romantic Friendship and Love Between Women from the Renaissance to the Present* (New York: William Morrow, 1981); Rothblum and Brehony, *Boston Marriages*; Terry Castle, *The Apparitional Lesbian: Female Homosexuality and Modern Culture* (New York: Columbia University Press, 1993), 11.

184. William Wordsworth, "To the Lady E.B. and the Hon. Miss P" (1824), *The Complete Poetical Works of William Wordsworth* (Cambridge, MA: The Riverside Press, 1904), 640.

185. Edith White, "Memories of Pioneer Childhood and Youth in French Corral and North San Juan, Nevada County, California. With a brief narrative of later life, told by Edith White, emigrant of 1859, to Linnie Marsh Wolfe, 1936," in Christiane Fischer, ed., *Let Them Speak for Themselves: Women in the American West 1849–1900* (Hamden, CT: Archon, 1977), 274–75.

186. Benita Eisler, ed., *The Lowell Offering: Writings by New England Mill Women (1840–1845)* (New York: W. W. Norton, 1997), 150.

187. Patricia Cooper and Norma Bradley Allen, *The Quilters: Women and Domestic Art: An Oral History* (Lubbock, TX: Texas Tech University Press, 1999), 29.

188. Marguerite Ickis, *The Standard Book of Quilt Making and Collecting* (New York: Dover Publications, 1959), 259.

189. Pamela A. Parmal and Jennifer M. Swope, eds., *Quilts and Color: The Pilgrim/Roy Collection* (Boston: MFA Publications, 2013), 82. See also Martha Schwendener, *The New York Times*, 20 Feb. 2015, C20.

190. Joanna L. Stratton, *Pioneer Women: Voices from the Kansas Frontier* (New York: Simon & Schuster, 2013), Kindle edition, ebook location 2840.

191. Jean V. Matthews, *The Rise of the New Woman: The Women's Movement in America 1875–1930* (Chicago: Ivan R. Dee, 2003), 17.

192. "In 1860, the most popular women's magazine of the era, *Godey's Lady's Book*, declared,

'The perfection of womanhood . . . is the wife and mother, the center of the family, that magnet that draws man to the domestic altar, that makes him a civilized being . . . The wife is truly the light of the home."' Quoted in Tiffany K. Wayne, *Women's Roles in Nineteenth-Century America* (Westport, CT: Greenwood Press, 2007), 1.

193. Virginia Woolf satirized the popular notion of the angel in the house in a 1931 lecture: "She was intensely sympathetic. She was immensely charming. She was utterly unselfish. She excelled in the difficult arts of family life. She sacrificed daily . . . Above all, she was pure." Virginia Woolf, "Professions for Women," in *The Death of the Moth and Other Essays* (Orlando, Florida: Harcourt Brace & Company, 1942), 235.

194. Carolyn J. Lawes, *Women and Reform in a New England Community, 1815–1860* (Lexington, KY: University Press of Kentucky, 2000), 64.

195. Ibid.

196. Erica Armstrong Dunbar, *A Fragile Freedom: African American Women and Emancipation in the Antebellum City* (New Haven, CT: Yale University Press, 2008), 61.

197. Ibid., 60.

198. Clifford M. Drury, "The Columbia Maternal Association," *Oregon Historical Quarterly* 39 (June 1938), quoted in Sandra Haarsager, *Organized Womanhood: Cultural Politics in the Pacific Northwest, 1840–1920* (Norman, OK: University of Oklahoma Press, 1997), 37.

199. William W. Fowler, *Woman on the American Frontier* (S. S. Scranton and Company, 1878; New York: Cosimo Inc., 2005).

200. "Athens of America Origin," Celebrate Boston, http://www.celebrateboston.com/culture/athens-of-america-origin.htm.

201. Bruce A. Ronda, *Elizabeth Palmer Peabody: A Reformer on Her Own Terms* (Cambridge, MA: Harvard University Press, 1999), 156.

202. Ibid., 187.

203. Margaret Fuller Ossoli, *Woman in the Nineteenth Century and Kindred Papers Relating to the Sphere, Condition and Duties, of Woman* (1844), ebook location 1243.

204. Margaret Fuller to Sophia Ripley, 27 Aug. 1839, "On the nature of the proposed Conversations," American Transcendentalism Web, http://transcendentalism-legacy.tamu.edu/authors/fuller/conversationsletter.html.

205. Joan von Mehren, *Minerva and the Muse: A Life of Margaret Fuller* (Amherst, MA: University of Massachusetts Press, 1995), 116.

206. Fuller, *Woman in the Nineteenth Century*, ebook location 1200.

207. Megan Marshall, *Margaret Fuller: A New American Life* (Boston: Houghton Mifflin Harcourt, 2013), 167.

208. Ibid., 181.

209. Ibid., 61, 92–93.

210. Charles Capper, *Margaret Fuller: An American Romantic Life, Volume II: The Public Years* (New York: Oxford University Press, 2007), 19.

211. Robert Hudspeth, ed., *The Letters of Margaret Fuller* (Cornell: Cornell University Press, 1987), 4:132.

212. "Besides the free Grammar School there were innumerable night schools; and most of the churches provided, by means of 'Social Circles,' opportunities for improvement." Daniel Dulany Addison, *Lucy Larcom: Life, Letters, and Diary* (Cambridge, MA: The Riverside Press, 1895), 7.

213. Lucy Larcom, *A New England Girlhood Outlined from Memory* (New York: Houghton Mifflin, 1889), 196.

214. Priscilla Murolo, *The Common Ground of Womanhood: Class, Gender, and Working Girls' Clubs 1884–1928* (Urbana: University of Illinois Press, 1997), 24.

215. Ibid., 158.

216. *Woman's Era* 1:19 (Dec. 1894), cited in Maude Thomas Jenkins, "The History of the Black Woman's Club Movement in America" (PhD diss., Columbia University Teacher's College, 1984), 51.

217. Maxine Seller, ed., *Immigrant Women* (Albany: State University of New York Press, 1994), 191.

218. This and the preceding quotations are from Elizabeth Cady Stanton, *Eighty Years and More: Reminiscences 1815–1897* (New York: Schocken Books, 1975), 162-94.

219. Jean V. Matthews, *The Rise of the New Woman: The Women's Movement in America, 1875–1930* (Chicago: Ivan R. Dee, 2003), 11.

220. William H. Chafe, *The Paradox of Change: American Women in the Twentieth Century* (New York: Oxford University Press, 1991), 99.

221. This and the preceding quotations, starting with the section "College Friends," are from Linda W. Rosenzweig, *Another Self: Middle-Class American Women and Their Friends in the Twentieth Century* (New York: New York University Press, 1999), 40, 41, 51–56.

222. The Hazel Traphagen and the Jette Johnson scrapbooks are in the Dept. of Special Collections, Stanford University Libraries.

223. Matthews, *The Rise of the New Woman*, 97.

224. Ibid., 98.

225. Mary Beth Norton and others, *A People, A Nation* (Boston: Houghton Mifflin, 2005), 512.

226. Chafe, *The Paradox of Change*, 13.

227. "The Story of a Sweatshop Girl: Sadie Frowne," *The Independent*, Sept. 25, 1902, quoted in *Plain Folk: The Life Stories of Undistinguished Americans*, ed. David M. Katzman and William M. Tuttle Jr. (Urbana: University of Illinois Press, 1982), 48–57.

228. Elizabeth Dutcher, "Budgets of the Triangle Fire Victims," *Life and Labor*, Sept. 1912, 266–67.

229. Thomas Jesse Jones, *Sociology of a New York City Block* (New York, 1904), 108–9, quoted in Kathy Peiss, "Gender Relations and Working-Class Leisure: New York City, 1880–1920," in *"To Toil the Livelong Day": America's Women at Work, 1780–1980*, ed. Carol Groneman and Mary Beth Norton (Ithaca, NY: Cornell University Press, 1987), 104.

230. This and the preceding quotations, beginning with "I can well recall . . . ," are from Jane Addams, *Twenty Years at Hull-House with Autobiographical Notes* (New York: The Macmillan Company, 1912).

231. Blanche Wiesen Cook, "Female Support Networks and Political Activism," in *A Heritage of Her Own*, ed. Nancy F. Cott and Elizabeth H. Pleck (New York: Simon & Schuster, 1979), 415–20.

232. Gioia Diliberto, *A Useful Woman: The Early Life of Jane Addams* (New York: Scribner, 1999); Jean Bethke Elshtain, *A Useful Woman: Jane Addams and the Dream of American Democracy* (New York: Basic Books, 2002); Louise W. Knight, *Citizen: Jane Addams and the Struggle for Democracy* (Chicago: University of Chicago Press, 2005).

233. This and the preceding three quotations are from Hilda Satt Polacheck, *I Came a Stranger: The Story of a Hull-House Girl* (Chicago: University of Chicago Press, 1989), 52, 167-68.

234. Quoted in Anne Firor Scott, *The Southern Lady: From Pedestal to Politics, 1830–1930* (Chicago: University of Chicago Press, 1970), 230.

235. Chafe, *The Paradox of Change*, 104.

236. This and the preceding quotes, beginning with "invisibly linked by the unique bond," are from Vera Brittain, *Testament of Friendship* (New York: Seaview Books, 1981), 84, 109-12, 114, 117, 145, 146, 2.

237. This and the preceding quotes are from Kristie Miller and Robert H. McGinnis, eds., *A Volume of Friendship: The Letters of Eleanor Roosevelt and Isabella Greenway, 1904–1953* (Tucson: Arizona Historical Society, 2009), 20, 190, 202, 261.

238. Blanche Wiesen Cook, *Eleanor Roosevelt, Volume I: 1884–1933* (New York: Viking, 1992), 292–93.

239. Doris Kearns Goodwin, *No Ordinary Time: Franklin and Eleanor Roosevelt: The Home Front in World War II* (New York: Simon & Schuster, 1994), 208.

240. Joseph P. Lash, *Love, Eleanor: Eleanor Roosevelt and Her Friends* (Garden City, NY: Doubleday & Company, Inc., 1982), 85.

241. Gail Collins, *America's Women: 400 Years of Dolls, Drudges, Helpmates, and Heroines* (New York: Harper Perennial, 2003), 362.

242. Lash, *Love, Eleanor*, 112.

243. Maurine H. Beasley, *Eleanor Roosevelt: Transformative First Lady* (Lawrence: University Press of Kansas, 2010), 136.

244. This and the preceding quotation are from Blanche Wiesen Cook, *Eleanor Roosevelt, Volume 2: The Defining Years, 1933–1938* (New York: Viking, 1999), 527–28, 533.

245. Lash, *Love, Eleanor*, 116–19.

246. Rodger Streitmatter, ed., *Empty Without You: The Intimate Letters of Eleanor Roosevelt and Lorena Hickok* (New York: Free Press, 1998), 16–22.

247. Lorena Hickok, *Reluctant First Lady* (New York: Dodd Mead, 1980).

248. Goodwin, *No Ordinary Time*, 122.

249. Ibid., 123.

250. The best-known of Lash's many books about Eleanor is *Eleanor and Franklin: The story of their relationship, based on Eleanor Roosevelt's private papers* (New York: Norton, 1, 1971 and 2, 1973), the first volume of which won him a Pulitzer Prize.

251. Allida M. Black, "Persistent Warrior: Eleanor Roosevelt and the Early Civil Rights Movement," in *Women in the Civil Rights Movement: Trailblazers and Torchbearers, 1941–1965*, ed. Vicki L. Crawford, Jacqueline Anne Rouse, and Barbara Woods (Bloomington: Indiana University Press, 1993), 243.

252. Allida M. Black, *Casting Her Own Shadow: Eleanor Roosevelt and the Shaping of Postwar Liberalism* (New York: Columbia University Press, 1996), 116.

253. For various takes on the couple as an historical construct, see *Inside the American Couple: New Thinking/New Challenges*, ed. Marilyn Yalom and Laura L. Carstensen (Berkeley: University of California Press, 2002).

254. Collins, *America's Women*, 362.

255. Diane Johnson, *Flyover Lives* (New York: Viking, 2014), 42–43.

256. Wallace Stegner, *Crossing to Safety* (New York: Modern Library, 2002), 277–78.

257. "Median Age at First Marriage by Sex: 1890–2010" (graph), United States Census Bureau, www.census.gov/hhes/socdemo/marriage/data/acs/ElliottetalPAA2012figs.pdf

258. Maxine Kumin, "Our Farm, My Inspiration," *American Scholar*, Winter 2014, 66.

259. Mirra Komarovsky, *Blue Collar Marriage* (New York: Vintage Books, 1962).

260. Carol Hanisch, "A Critique of the Miss America Protest" (1968), in *Women's America: Refocusing the past*, ed. Linda K. Kerber and Jane Sherron De Hart, 577.

261. ibid.

262. Carol P. Christ and Judith Plaskow, eds., *Womanspirit Rising: A Feminist Reader in Religion* (New York: Harper & Row, 1979), 204.

263. This and the preceding quotations are from Carolyn See, "Best Friend, My Wellspring in the Wilderness!" in *Between Friends: Writing Women Celebrate Friendship*, ed. Mickey Pearlman (New York: Houghton Mifflin Company, 1994), 56–73.

264. Nikki Giovanni, *Gemini* (New York: William Morrow, 1971), 37.

265. Toni Morrison, *Sula* (New York: Knopf, 1974), 5.

266. Claudia Tate, ed., *Black Women Writers at Work* (New York: Continuum, 1983), 118.

267. Gloria Naylor, *The Women of Brewster Place* (New York: Viking, 1982), 103–4.

268. Alice Walker, *The Color Purple* (New York: Pocket Books, 1985), 42.

269. Margaret Talbot, "Girls Just Want to Be Mean," *New York Times Magazine*, February 24, 2002, www.nytimes.com/2002/02/24/magazine/girls-just-want-to-be-mean.html.

270. Rebecca Raber, "The 10 Best Female Friendships in Television History," TakePart, www.takepart.com/photos/10-best-female-friendships-television-history/-10-the-mary-tyler-moore-show-mary-and-rhoda.

271. Dave Itzkoff, "Taking an Express to Cult Fame," *The New York Times*, January 13, 2015.

272. Nick Paumgarten, "Id Girls," *New Yorker*, June 23, 2014, 40.

273. This and the preceding quotations are from Chrisena Coleman, *Just Between Girlfriends: African-American Women Celebrate Friendship* (New York: Simon & Schuster, 1998), 61, 68–70.

274. N. Lynne Westfield, *Dear Sisters: A Womanist Practice of Hospitality* (Cleveland: Pilgrim Press, 2001), 65.

275. *The Mother's Study Club: The First Century 1914–2014* (Concord, NH: Town & Country, 2013), 200–7.

276. Eighty-five percent of adults in the United States and nearly 40 percent of the world's population have access to the Internet. "Internet users per 100 inhabitants 2006–2013" (table), International Telecommunications Union, in "Global Internet Usage," Wikipedia, en.wikipedia.org/wiki/Global_Internet_usage (accessed June 3, 2013). See also the 2013 and 2014 Pew Internet and American Life Project, www.pewinternet.org.

277. Online, women as a group live up to their reputation as the more communicative gender. For instance, 71 percent of women online use social networking sites, versus 62 percent of men. And each month, forty million more women than men visit Twitter. Men tend to dominate technical sites, but the big sites geared toward general communication and sharing are distributed roughly 40 percent male to 60 percent female. Women drive content at Pinterest (79 percent), Goodreads (70 percent), and Blogger (66 percent). "Report: Social network demographics in 2012," Pingdom, royal.pingdom.com/2012/08/21/report-social-network-demographics-in-2012/.

278. Jenna Goudreau, "What Men and Women Are Doing on Facebook," *Forbes*, www.forbes.com/2010/04/26/popular-social-networking-sites-forbes-woman-time-facebook-twitter.html.

279. http://www.tericase.com/?p=183#more-183.

280. Kamy Wicoff (founder, SheWrites.com), interview with the author, 19 Aug. 2013.

281. Ibid.

282. Sponsored by Magasin III Museum and Foundation for Contemporary Art, Stockholm.

283. Miranda July, We Think Alone, http://wethinkalone.com/about/.

284. Joe Navarro with Marvin Karlins, *What Every Body Is Saying: An Ex-FBI Agent's Guide to Speed-Reading People* (New York: William Morrow, 2008), Kindle e-book, locations 149–93.

285. A corollary is that young players of violent video and online games may fail to grasp the difference between pretend aggression and the real thing.

286. Teri Evans, "Reaping Success Through Stranger 'Meetups," *Wall Street Journal* online, 21 Nov. 2010. http://www.wsj.com/articles/SB10001424052748704170404575624733792905708.

287. Rebecca Tuhus-Dubrow, "Women Can Connect, Click by Click," *New York Times*, 13 July 2012.

288. Shasta Nelson (founder, GirlFriendCircles.com), in discussion with the author, 5 Sept. 2013.

289. Shoshana K., "A Success Story: Shoshana Is Making Friends in L.A.," Shasta's Friendship Blog, GirlFriendCircles, girlfriendcircles.com/blog/index.php/2013/07/a-success-story-girlfriendcircles-make-friends-la/.

290. "The labor force participation rate—the percent of the population working or looking for work—for all mothers with children under age 18 was 69.9 percent in 2013." "Employment Characteristics of Families Summary," United States Bureau of Labor Statistics, 25 Apr. 2014, http://www.bls.gov/news.release/famee.nr0.htm.

291. A coinage attributed to Thomas Carlyle, who described Thomas Malthus's eighteenth-century economic theory—that population growth would inevitably exceed the food supply—as "dismal." Thomas Carlyle, *Chartism*, 2nd ed. (London: James Fraser, 1840), 109.

292. Nicholas A. Christakis, *Connected: The Surprising Power of Our Social Networks* (New York: Back Bay Books, 2009), 18.

293. "Since market rationality is central to neoliberalism, the market becomes spread across our lives. Not only our economic but also our political, social, and personal relationships all become markets." Todd May, *Friendship in an Age of Economics: Resisting the Forces of Neoliberalism* (Lanham, MD: Lexington Books, 2012), 30.

294. May, "Friendship in an Age of Economics," opinionator.blogs.nytimes.com/2010/07/04.

295. Ken Auletta, "A Woman's Place: Can Sheryl Sandberg Upend Silicon Valley's Male-Dominated Culture?," *New Yorker*, 11 July 2011.

296. Arlie Hochschild, *The Outsourced Self: Intimate Lives in Market Times* (New York: Metropolitan Books, 2012), 8–9.

297. Ibid., 195–96.

298. Shelley Taylor, *The Tending Instinct*, 94.

299. Leah Busque, talk at ecorner, Stanford University's Entrepreneur's Corner, May, 2014. http://ecorner.stanford.edu/authorMaterialInfo.html?mid=3349

300. As told to author. Names have been changed.

301. Revolution Foods, which seeks to fight the obesity epidemic by providing healthy meals to schoolchildren, made CNNMoney's list of the one hundred fastest-growing inner-city companies in 2012. *Inc.* magazine placed the company sixth on its list of most innovative food companies, and its founders "made *Fortune*'s '40 Under 40' list of 'young hotshots who are rocking businesses.'" Kim Girard, "Expanding the Menu," *BerkeleyHaas* magazine, Fall 2013, 11.

302. Sylvia Ann Hewlett, "Mentors Are Good. Sponsors Are Better," *New York Times*, April 13, 2013.

303. Ibid.

304. "World Bank Group: Working to End Extreme Poverty and Hunger," World Bank, http://www.worldbank.org/mdgs/poverty_hunger.html.

305. "Approach to Microfinance," Opportunity International, http://opportunity.org/what-we-do/microfinance.

306. Ibid.

307. "New England Centenarian Study," Boston University School of Medicine, http://www.bumc.bu.edu/centenarian.

308. Karen M. Bush, Louise S. Machinist, and Jean McQuillin, *My House Our House: Living Far Better for Far Less in a Cooperative Household* (Pittsburgh: St. Lynn's Press, 2013), 56–58.

309. Ibid.

310. Sarah Mahoney, "The New Housemates," *AARP The Magazine*, July 2007, www.aarp.org/home-garden/housing/info-2007/the_new_housemates.

311. See Bush, Machinist, and McQuillin, *My House Our House*, for a helpful resource list and great advice about cohousing. A good online clearinghouse for cohousing information is Women for Living in Community, www.womenlivingincommunity.com.

312. Ibid.

313. For a discussion of sex segregation during childhood socialization, see Eleanor E. Maccoby, *The Two Sexes: Growing Up Apart, Coming Together* (Cambridge, MA: Harvard University Press, 1998), 118–52.

314. Sheikh Assim L. Alhakeem, "Is Friendship Between Man and Woman Allowed in Islam?," www.youtube.com/watch?v=Z8hXwIQG2sw.

315. "Children tease each other for showing signs of 'liking' or 'loving' a child of the other sex, and presumably this teasing cuts off approaches that might otherwise be made to an other-sex child." Maccoby, *The Two Sexes*, 289.

316. Adrian F. Ward, "Men and Women Can't Be 'Just Friends,'" *Scientific American*, Oct. 23, 2012, http://www.scientificamerican.com/article/men-and-women-cant-be-just-friends/.

317. Ibid., 169.

318. J. Bradley Blankenship, "Gender-Blind Housing: College Men and Women Living Together," Kinsey Confidential, 20 Sept. 2011, http://kinseyconfidential.org/genderblind-housing-college-men-women-living/.

319. Georgia Wisdom, "10 Rules for Friends with Benefits," Thought Catalog, 8 Feb. 2013, http://thoughtcatalog.com/georgia-wisdom/2013/02/10-rules-for-friends-with-benefits/.

320. "[In 2000], researchers had 309 college students (roughly half of either sex) complete a survey concerning sexual activity in nonromantic cross-sex friendships . . . 51% of the

respondents had sex with a cross-sex friend whom they had no intention of dating . . . 44% of those who engaged in sexual activity eventually transformed the relationship into a romantic one." Michael Monsour, *Women and Men as Friends: Relationships Across the Life Span in the 21st Century* (Mahwah, NJ: Lawrence Erlbaum Associates, 2002), 138–39.

321. Conversation with author, Oct. 2013.

322. Sylvia Ann Hewlett, "As a Leader, Create a Culture of Sponsorship," *Harvard Business Review*, 8 Oct. 2013, https://hbr.org/2013/10/as-a-leader-create-a-culture-of-sponsorship.

323. Thomas Rogers, "Ladies: I'm Not Your Gay Boyfriend," *Salon*, 18 Aug. 2009, http://www.salon.com/2009/08/18/rogers_fag_hag/.

324. *National Health Statistics Reports* Number 49, 22 Mar. 2012.

325. Linda Dodge Reid, "The Family We Choose," *Stanford* magazine, September/October 2013.

326. Once upon a time, Americans thought French men were peculiar because they kissed each other on both cheeks. Greek men walking down the street holding hands caused American tourists to lift their eyebrows. Not anymore. We are getting used to seeing the rituals of kissing and hugging among men—and in places where they have never been before. For the first time in history, Chinese family members and friends are adopting the Western-style hug after an age-long prohibition against the practice. Consider this May 9, 2014, article from *The New York Times*, with the heading "Cautious Chinese Gain Comfort with Hugs":

   [At the Beijing airport] a younger Chinese couple greeted an older couple with hugs. The women embraced first, but the young man followed by squeezing the older man, stiffly. Here questions of age, as well as gender, enter into the embrace, with women making the overtures and the men (awkwardly) following their lead.

# ACKNOWLEDGMENTS

W E ARE GRATEFUL FOR THE help of many friendly people who contributed to this project. At Stanford University, we thank senior scholars Susan Groag Bell, Edith Gelles, and Karen Offen from the Michelle R. Clayman Institute for Gender Research; labor economist and Professor Emerita Myra Strober; Professor Emerita of English Barbara Gelpi; Professor Jane Shaw, Dean of Religious Life; and Mattie Taormina from the Special Collections of the Stanford University Libraries. Judith C. Brown, Professor Emerita of History at Wesleyan University, and historian Allida Black gave helpful advice for specific chapters. Social media insights were provided by Kamy Wicoff, founder of SheWrites.com; Shasta

Nelson, founder of GirlFriendCircles.com; and Ivory Madison, founder of RedRoom.com. Business consultant Anne Litwin shared her perspective. Anna Paustenbach provided insightful comments. Irvin Yalom read and critiqued the entire manuscript. Paul, Julia, and Gracie Brown enlivened our understanding of friendship in the twenty-first century. Special thanks to Gail Winston, our editor at HarperCollins, and to Sandra Dijkstra, our literary agent, who shepherded this book to completion. Of course, the generous women who shared their stories of friendship are too numerous to name.

# INDEX

## ABOUT THE AUTHORS

MARILYN YALOM is a former professor of French and a senior scholar at the Clayman Institute for Gender Research at Stanford University. She is the author of widely acclaimed books such as *A History of the Breast*, *A History of the Wife*, *Birth of the Chess Queen*, and, most recently, *How the French Invented Love*. She lives in Palo Alto, California, with her husband, psychiatrist and author Irvin D. Yalom.

THERESA DONOVAN BROWN is an award-winning author of both fiction and nonfiction. She holds a BA from Stanford University and an MBA from the Haas School of Business at the University of California, Berkeley. Her background writing policy-level speeches for global economic leaders, trading securities, and running a financial communications firm informs her insights into the cultural history of friendship.

# BOOKS BY MARILYN YALOM

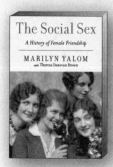

## THE SOCIAL SEX
**A History of Female Friendship**
with Theresa Donovan Brown

Available in Paperback and eBook

Surveying history, literature, philosophy, religion, and pop culture, *The Social Sex* demonstrates how women were able to co-opt the public face of friendship throughout the years. Chronicling shifting attitudes toward friendship, Yalom and Brown reveal how the concept of female friendship has been inextricably linked to the larger social and cultural movements that have defined human history.

## HOW THE FRENCH INVENTED LOVE

Available in Paperback and eBook

"*How the French Invented Love* is absolutely marvelous, so lively and learned. . . . Marilyn Yalom's book is a distinguished contribution to our experience of a great literature, as well as an endearing memoir."
—Diane Johnson, author of *Lulu in Marrakech* and *Le Divorce*

## BIRTH OF THE CHESS QUEEN

Available in Paperback and eBook

Everyone knows that the queen is the most dominant piece in chess, but few people know that the game existed for five hundred years without her. It wasn't until chess became a popular pastime for European royals that the queen was born.

"A delectably readable volume." —*People*

## A HISTORY OF THE WIFE

Available in Paperback and eBook

For any woman who is, has been, or ever will be married, this intellectually vigorous and gripping historical analysis of marriage sheds new light on an institution most people take for granted.

"A useful overview of women's changing roles in marriage and society." —*Kirkus*